I0794035

RUSSIAN AIRCRAFT OF WORLD WAR II 1939–45

RUSSIAN AIRCRAFT OF WORLD WAR II
1939–45

EDWARD WARD

amber
BOOKS

First published in 2021

Reprinted in 2025

Copyright © 2021 Amber Books Ltd

All rights reserved. No part of this publication may be reproduced, stored in a retrieval system, or transmitted in any form or by any means, electronic, mechanical, photocopying, recording, or otherwise, without prior written permission of the copyright holder.

Published by Amber Books Ltd
United House
London N7 9DP
United Kingdom
www.amberbooks.co.uk
Facebook: amberbooks
Instagram: amberbooksltd
Twitter: @amberbooks
Pinterest: amberbooksltd

ISBN: 978-1-83886-083-7

Editor: Michael Spilling
Designer: Mark Batley
Picture research: Terry Forshaw

Printed in China

Contents

Introduction

On 22 June 1941, Hitler unleashed Operation Barbarossa on an unsuspecting Soviet Union, beginning what the Russians referred to as the 'Great Patriotic War' and pitting the vaunted *Luftwaffe* against the world's largest air force, the *Voyenno-Vozdushnye Sily* ('Military Air Forces'), or VVS. By the end of the day the *Luftwaffe* claimed 1489 Soviet aircraft destroyed on the ground. This figure was initially believed to be inflated, but in fact it was an underestimation – over 2000 wrecks were counted on captured airfields, a figure confirmed by Soviet sources.

Unfortunately for the VVS, the German strike came at a time of transition. Aircraft like the Yak-1, Ilyushin Il-2 and Petlyakov Pe-2 were of world standard but few in number. Years had passed in which design had been allowed to stagnate and the majority of combat aircraft were relatively old designs, such as the Polikarpov I-16 and Tupolev SB-2.

This situation was compounded by Stalin's overwhelming paranoia – around 6000 experienced officers were purged from 1937 onwards, not least Generals Pavel Rychagov and Yakov Smushkevich, both commanders of the VVS who had fought in Spain and had achieved air supremacy over the Japanese at Khalkhin Gol in 1939. Both were tortured and killed. By June 1941, 90 per cent of unit leaders had been in position for six months or less, and inexperience led to many costly errors. One such mistake was the failure to

Bearing the inscription 'Little Theatre: Front' (i.e. donated by Moscow's Little Theatre for the Front), these Yak-9s were destined for operations in Byelorussia.

Comparatively unknown outside the land of its birth, the Pe-2 was the Soviet counterpart of the Mosquito, but made of metal. Indeed, it was built in considerably greater numbers than the British aircraft, and for as many different types of mission. The Klimov VK-105 12-cylinder, liquid-cooled, inline engine that powered the Pe-2 was a derivative of the VK-100, a licence-built version of the Hispano 12Y of the 1930s.

properly disperse aircraft. Most were neatly lined up, aiding German strafing attacks. The VVS flyers who managed to take the fight to the Germans were generally outclassed; pilot training time had been cut in March 1940 and again in February 1941, resulting in a disastrous drop in pilot capability. Meanwhile, many designers were incarcerated: some of the USSR's finest aircraft – such as the Petlyakov Pe-2 and Tupolev Tu-2 – were designed in prison.

Industrial strength

But Soviet recovery was swift and impressive. One of the major blunders of German intelligence was a serious underestimation of Soviet industrial might. The *Luftwaffe* believed Soviet production capability was 600 aircraft per month in 1939, while the real figure was between 3500 and 4000. It was also known that there was a plan to move aircraft factories eastwards beyond the Urals, but it was believed by the Germans to be impossible to carry out. In the event this is exactly what did happen, putting Soviet industry beyond the range of any German air attack.

Loss of output at this critical time was balanced by the flow of lend-lease American and British aircraft, such as the Hawker Hurricane, Curtiss P-40 Warhawk, Douglas A-20 and the popular Bell P-39 Airacobra.

Domestic aircraft would improve qualitatively: by 1945, aircraft such as the Yak-3 and Lavochkin La-7 were amongst the world's best and available in quantity. Similarly, after 1941 the quality of training continually increased and vast numbers of combat aircrew went into action. Uniquely amongst the fighting powers, this included around 3000 women.

Tactical force

Like the *Luftwaffe*, the VVS was primarily a tactical force, intended to support the Red Army in the field. It is not a coincidence that the most-produced combat aircraft of all time, the Il-2, was a ground attack aircraft. The majority of Soviet bombers were twin engine machines like their German counterparts, the Tu-2 being broadly analogous to the similarly capable Junkers Ju 88, for example. Neither nation achieved much in the realm of strategic bombing, which in the Soviet case is surprising, since with the Petlyakov Pe-8 they possessed a world-class, long-range bomber.

Although the war on the Eastern Front was primarily a land campaign, Soviet naval aviation played an important part, too. Many units were manned by naval personnel and long-range maritime patrols, anti-shipping and anti-submarine missions were flown from the Arctic to the Pacific.

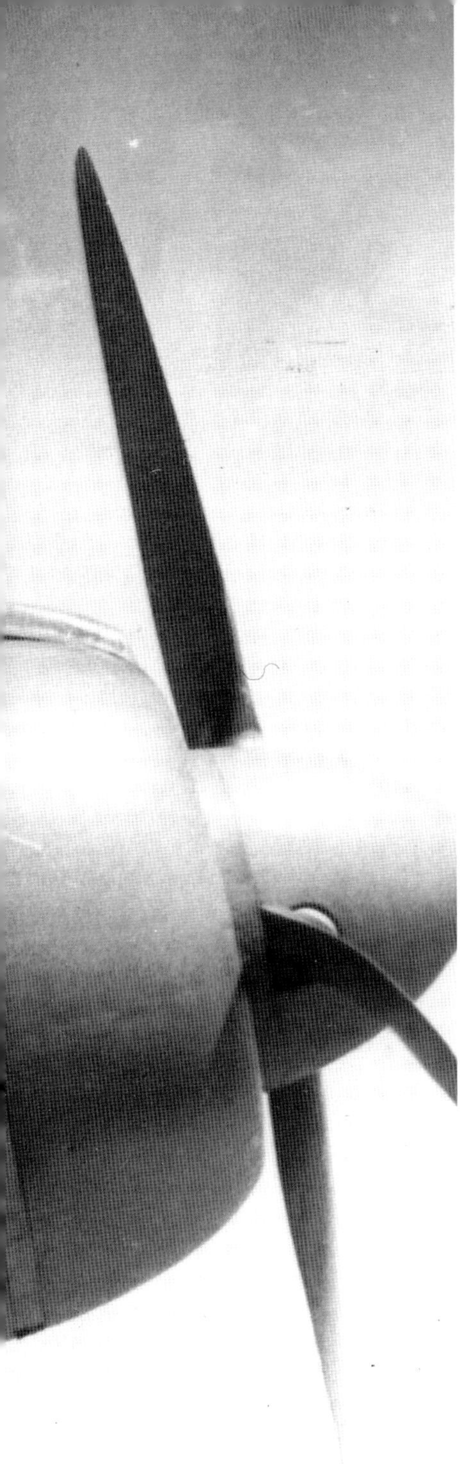

FIGHTER AIRCRAFT

Soviet fighter design had slipped behind world standard when Hitler launched Operation Barbarossa on 22 June 1941 and designers responded with frantic but ultimately successful efforts to catch up. The qualitative gap – especially during the dark days of late 1941 and early 1942 – was filled by vast quantities of foreign aircraft. By 1945 the Soviet Union was once again building fighters equal to any in the world.

This chapter includes the following aircraft:

- Polikarpov I-15
- Polikarpov I-153
- Polikarpov I-16
- Polikarpov I-185
- LaGG-3
- MiG-1
- MiG-3
- Petlyakov Pe-3
- Yakovlev fighters
- Lavochkin La–5
- Lavochkin La–7
- Bereznyak-Isayev BI-1
- Curtiss P-40
- Hawker Hurricane
- Bell P-39 Airacobra
- Bell P-63 Kingcobra
- Supermarine Spitfire
- P-47 Thunderbolt

A Soviet pilot poses standing on the wing of a Lavochkin La-5FN. The aircraft was designed to take on Messerschmitt Bf 109s at low altitudes.

Polikarpov I-15

The manoeuvrable I-15 proved highly effective during the Spanish Civil War but by the time of the German invasion the zenith of its development was long past.

Nikolai Polikarpov's rotund I-15 first flew in 1933 in the hands of famed test pilot Valeri Chkalov. A replacement for Polikarpov's own successful I-5, the new aircraft demonstrated exceptional manoeuvrability and climb, attaining 5000m (16,250ft) in just over six minutes. Key to its impressive performance was its American Wright R-1820 Cyclone engine, which Shvetsov would produce under licence as the M-25 although all early I-15s were powered by imported engines. Armament was provided by four 7.62mm (0.3in) PV-1 machine guns.

Effective combat service

Entering service in 1934, the I-15 proved generally popular with pilots although reservations concerning stability and the lack of pilot visibility afforded by the gulled arrangement of the upper wing led to the development of a version with conventional cabane struts and centre section. Known as the I-15bis (or I-152), the redesigned aircraft possessed virtually identical performance to its predecessor and entered production in 1937. By this time the I-15 had seen extensive operational service in Spain, its first combat occurring on 4 November 1936 when I-15s shot down two Ju 52/3ms and two Fiat CR.32 fighters without loss. Against the principal Nationalist fighters, the I-15 proved markedly superior to the Heinkel He 51 and broadly equivalent to the CR.32. However, as more advanced monoplane fighters appeared the I-15 was increasingly sidelined in speed although its manoeuvrability remained unmatched.

Ground-attack role

After seeing action in the Winter War and over Khalkhin Gol against the Finnish and Japanese respectively, by June 1941 around 1000 I-15 and I-15bis remained in front line service. By now totally obsolete, the I-15 suffered heavy losses in air combat against the Germans, although in the hands of an experienced pilot its sparkling agility could still cause problems for opposing fighters. Although lacking armour and vulnerable even to small-arms fire, the I-15 enjoyed greater success as a ground-attack aircraft despite only carrying a modest 50kg (110lb) bombload or up to six RS-82 rockets. By the end of 1942 all surviving examples were employed in secondary roles.

Polikarpov I-15

Weight (Maximum take-off) 1689kg (3724lbs)

Dimensions Length: 6.10m (20ft), Wingspan: 9.75m (32ft), Height: 2.20m (7ft 3in)

Powerplant One 520kW (700hp) Shvetsov M-25 9-cylinder air-cooled radial engine

Speed 367km/h (228mph)

Range 510km (320 miles)

Ceiling 9800m (32,200ft)

Crew 1

Armament Four 7.62mm (0.3in) PV-1 machine guns; up to 50kg (110lbs) bombload

Polikarpov I-15bis

Serving with 70 IAP during the Battle of Khalkhin Gol against the Japanese, this I-15bis was based on the border of Mongolia and the puppet state of Manchukuo in the summer of 1939.

Polikarpov I-153

The most formidable biplane fighter to see widespread service, the Polikarpov I-153 took development of this class of aircraft to its limit.

Even as other nations were coming to view the biplane layout as passé, experience gained fighting the nimble Fiat CR.32 over Spain convinced Russian planners that biplane fighters warranted further development and the Polikarpov I-153 was the direct result. A development of the I-15bis, the I-153 was intended to maximize combat performance without sacrificing the exemplary manoeuvrability of the older aircraft, the most obvious external change being the replacement of the fixed landing gear with a retractable undercarriage.

First flown in August 1938, the I-153 reverted to the gull-wing centre section of the I-15 to cut down drag, inspiring its *Chaika* (seagull). nickname. Four ShKAS machine guns were fitted of the same 7.62mm (0.3in) calibre as the I-15's PV-1s but possessing more than double the rate of fire.

The M-62 engine intended for the new aircraft was not yet available so a developed version of the I-15's M-25 was employed developing 560kW (750hp), and the I-153 demonstrated performance well in advance of the earlier aircraft, the only significant shortcoming being a difficult spin recovery. As the aircraft was found to be reluctant to enter a spin in the first place this was deemed a relatively insignificant drawback, therefore production was initiated at Zavod 156 in Moscow and the aircraft was introduced into service in the spring of 1939.

I-153 versus I-153

Almost immediately the I-153 was committed to combat against the Japanese in the battles of Khalkhin Gol (11 May 1939–16 September 1939). Soviet I-16s were struggling to cope with the Japanese Army's Nakajima Ki 27, quite possibly the most agile monoplane fighter ever built, and I-153s were rushed to the front by rail to redress the balance. The two aircraft were well matched and the success

Polikarpov I-153

Weight (Maximum take-off) 2110kg (4652lbs)

Dimensions Length: 6.17m (20ft 3in), Wingspan: 10m (32ft 10in), Height: 2.8m (9ft 2in)

Powerplant One 745kW (1000hp) Shvetsov M-62 9-cylinder air-cooled radial piston engine

Speed 297km/h (185mph)

Range 470km (290 miles)

Ceiling 10,700m (35,100ft)

Crew 1

Armament Four 7.62mm (0.3in) ShKAS machine guns; eight RS-82 rockets

Polikarpov I-153

Based at Lavansaari on the Gulf of Finland, this I-153 of the 71 IAP of the Baltic Fleet was flown by the unit's commanding officer, Petr Biskup. Biskup scored four of his five victories in the I-153.

of the I-153 in this theatre appeared to vindicate the decision to develop a modern biplane fighter. During the Winter War of 1940–41, the I-153 performed adequately enough against the relatively poorly equipped Finns, themselves enthusiastic users of the type who were operating some 22 captured examples. On one occasion three Finnish *Chaikas* were bounced by two Russian I-153s, resulting in the loss of one Finnish aircraft, the I-153 becoming in the process one of very few aircraft to have fought itself.

Bullish production

By this time it was clear that future fighter development definitively favoured the monoplane, but production at two factories was in full swing and deliveries during the 18 month production run of the I-153 averaged a colossal 48 per week. From June 1939 the M-62 engine substituted the M-25 and powered most production aircraft. Featuring a two-speed supercharger and delivering 745kW (1000hp) at take-off, performance with the new engine remained unchanged at sea level but climb and speed at altitude were usefully improved. The increased power was accompanied by greater fuel consumption resulting in reduced range so provision was made for two underwing 100 litre (22 gallon) fuel tanks to compensate. Eventually, the final 409 aircraft received the higher compression M-63 engine offering 821kW (1100hp) at take-off.

Armament was improved with the adoption of four 12.7mm (.50in) BS machine guns, although eight I-153s were fitted with two 20mm (0.78in) ShVAK cannons and designated I-153P for *Pushechny* 'cannon'. Gun armament was augmented with racks for up to eight RS-82 rockets.

As well as testing a pressurized cabin, which was successful but unacceptably heavy for service, the I-153 was utilized in trials of the DM-4 ramjet, a pair of these motors boosting speed to 440km/h (273mph), although the detrimental effect on performance when these were not running outweighed the performance increase bestowed by this radical powerplant.

End of the line

The last of the 3437 I-153s built rolled off the Zavod 156 line in late 1940, followed closely by the first

Soviet aircrew study a map before operations, a Polikarpov I-153 fighter parked in the background.

Polikarpov I-153

Originally silver, 'White 19' was roughly brush-painted green whilst serving on the Karelian Front during the autumn of 1941. This aircraft was later repaired and pressed into service as IT-20 by the Finns.

Polikarpov I-153

Veteran of the 1939–40 Winter War, Konstantin Solovev of 71 IAP flew this I-153 at Suomenlahti in Finland. Solovev shot down a Fokker D.XXI in this aircraft on 27 March 1942.

Polikarpov I-153

'Black 69' wears the standard silver and light grey livery in which most prewar I-153s left the factory. This aircraft was flying with an unknown unit when it was captured following a forced landing during the summer of 1941.

MiG-1s, yet despite its replacement on the production line, I-153s made up nearly half the front line Soviet fighter complement when Operation Barbarossa was launched on 22 June 1941.

Against the *Luftwaffe*, arguably the best equipped and most experienced air force in the world, the I-153 stood little chance, although its phenomenal agility – if well handled – made it a difficult target for the Bf 109 and a few exceptional Soviet pilots would score their first victories with the *Chaika*. Increasingly employed in the close air-support role, the I-153 soldiered on until mid-1943 in front line units, although the final users were the Finns, who operated the I-153 until the end of hostilities.

Polikarpov I-153

Weight (Maximum take-off) 2110kg (4652lbs)
Dimensions Length: 6.17m (20ft 3in), Wingspan: 10m (32ft 10in), Height: 2.8m (9ft 2in)
Powerplant One 745kW (1000hp) Shvetsov M-62 9-cylinder air-cooled radial piston engine
Speed 297km/h (185mph)
Range 470km (290 miles)
Ceiling 10,700m (35,100ft)
Crew 1
Armament Four 7.62mm (0.3in) ShKAS machine guns; eight RS-82 rockets

Polikarpov I-16

One of the most significant combat aircraft in history, Nikolai Polikarpov's I-16 was the world's first cantilever monoplane fighter with enclosed cockpit and a retractable undercarriage.

Development of the I-16 was incredibly rapid. Design work began in March 1933, leading to the first flight on 31 December. Intended to be powered by the Shvetsov M-25, a licence-built Wright Cyclone, delays with the new engine meant the first prototype and initial production aircraft utilized the readily available but lower powered Shvetsov M-22 (a licence-built Bristol Jupiter). These were designated I-16 Type 4, there being no Type 1, 2 or 3, although confusingly initial production with both M-22 and M-25 engines shared the Type 4 designation.

Multiple design problems

Trials quickly revealed the aircraft was tricky to fly. The abbreviated fuselage and massive ailerons rendered the aircraft extremely sensitive to control movements on all three axes and it was susceptible to stalling and spinning with little provocation. The rotund fuselage made the aircraft difficult for pilots to enter and exit and early aircraft featured a claustrophobic forward-sliding enclosed canopy that gradually became opaque when exposed to sunlight. Furthermore, the hand-operated undercarriage required a physically demanding 44 turns of a handle mounted on the starboard side of the cockpit to fully retract.

However, the performance of the new aircraft, even with the low powered engine, was at the forefront of international fighter design, rate of climb and ceiling being considered the best in its class. When fitted with the M-25 the I-16 was comfortably the fastest fighter in the world. Production built up rapidly and during 1935 a switch was made to the new Type 5 featuring a more powerful M-25A engine. The Type 5 was the first I-16 variant to see combat when on 9 November 1936 12 Spanish I-16s intercepted a Nationalist bombing raid on Madrid, claiming four aircraft shot down. Over Spain the I-16 wrested control of the air from the rebels, forcing Nationalist bombers to operate at night and proving initially superior to all other fighter aircraft in the theatre. Combat experience did reveal some concerns, the greatest of which was armament. Polikarpov, intent on minimizing the weight of the new fighter, designed the I-16 with only two

Polikarpov I-16 Type 28

An I-16 Type 28 of an unknown unit pictured during the summer of 1942. The cannon armed Type 28 was built in relatively small numbers.

Polikarpov I-16 Type 28

Weight (Maximum take-off) 1988kg (4393lbs)

Dimensions Length: 6.1m (20ft 1in), Wingspan: 9m (29ft 6in), Height: 2.25m (7ft 5in)

Powerplant One 820kW (1100hp) Shvetsov M-63 9-cylinder supercharged air-cooled radial piston engine

Speed 463km/h (288mph)

Range 440km (270 miles)

Ceiling 9700m (31,800ft)

Crew 1

Armament Two 7.62mm (0.3in) ShKAS machine guns in upper cowling, two 20mm (0.78in) ShVAK cannons in the wings; six unguided RS-82 rockets or up to 500kg (1100lbs) of bombs

Polikarpov I-16

Flying in the Murmansk area during August 1941, this I-16 flew with the 72 IAP of the Northern Fleet Air Force. '*Za CCCP!*' translates as 'For the USSR!'

Polikarpov I-16 Type 18

Ace Boris Safanov flew this I-16 Type 18 over Murmansk in the summer of 1941. '*Za Stalina!*' translates as 'For Stalin!'. Safanov's I-16 was emblazoned with the slogan '*Smert Fashistam!*' ('Death to Fascists!) on the starboard side.

Polikarpov I-16 Type 18

Weight (Maximum take-off) 1830kg (4044lbs)

Dimensions Length: 6m (19ft 8in), Wingspan: 9m (29ft 6in), Height: 2.57m (8ft 5in)

Powerplant One 746kW (1000hp) Shvetsov M-62 9-cylinder radial piston engine

Speed 463km/h (288mph)

Range 805km (500 miles)

Ceiling 9700m (31,800ft)

Crew 1

Armament Four 7.62mm (0.3in) ShKAS machine guns in upper cowling and wings; six unguided RS-82 rockets or up to 500kg (1100lbs) of bombs

7.62mm (0.3in) ShKAS machine guns that were considered inadequate.

In 1937 production shifted to the next major variant, the Type 10, which featured four ShKAS weapons: two in the nose and two in the wings. The Type 10 benefitted from a strengthened structure and the more powerful M-25V engine as well as a fixed windscreen and rearward sliding canopy, although in service the canopy was usually discarded. An engine change to the Shvetsov M-62 resulted in the Type 18, in which strenuous efforts had been made to make the aircraft easier to fly, all further I-16s being notably less 'twitchy' than their forebears. This and the more powerful Type 24 of 1940 were built in vast numbers in the lead up to the Great Patriotic War, during which time I-16s were committed to combat over China and Mongolia against the Japanese and against the Finns in Karelia.

Polikarpov I-16 Type 6

This aircraft flew as part of the Spanish Republican Air Arm in 1938. The first machines to arrive in Spain went into battle on 15 November 1936, providing air cover for a Republican offensive against Nationalist forces advancing on Valdemoro, Sesena and Equivias. The total number of I-16s delivered to Spain between 1936 and 1938 was 276, of which 112 were lost in air-to-air combat. A further 42 I-16s were built in Spain, though only four were completed before the end of hostilities.

Tail Markings
The distinctive 'Popeye' markings on the tail of this I-16 signifies that it is an aircraft of the 4th Escuadrilla.

Polikarpov I-16 Type 6
Weight (Maximum take-off) 1508kg (3325lbs)
Dimensions Length: 5.99m (19ft 8in), Wingspan: 9m (29ft 6in), Height: 2.25m (7ft 5in)
Powerplant One 537kW (730hp) Shvetsov M-25A 9-cylinder radial piston engine
Speed 445km/h (277mph)
Range 540km (336 miles)
Ceiling 9100m (29,856ft)
Crew 1
Armament Three 7.62mm (0.3in) ShKAS machine guns

Cockpit Gun Sight
Early types featured a tubular gun sight and an enclosed canopy. Unusually, the entire canopy and windscreen assembly slid forwards to open and backwards to close.

Engine
The I-16 Type 6 was powered by a 537kW (730hp) Shvetsov M-25A nine-cylinder engine. Later types were upgraded to the more powerful Shvetsov M-62.

Armament
Before the development of the Type 6, I-16s only carried two 7.62mm (0.3in) ShKAS machine guns. The Type 6 added a third ShKAS in the fuselage.

It became apparent with each successive conflict that the I-16's performance was lagging further behind world standard. By the time of the German assault the I-16 was obsolescent but the I-16 remained the most commonly encountered Soviet fighter until the end of 1942. As a bomber destroyer the Polikarpov struggled: the Junkers Ju 88 was faster than the I-16, and although it could catch the He 111, it was difficult to shoot down the sturdy Heinkel.

In an effort to address this deficiency the final major variant, the type 29, augmented the two nose-mounted ShKAS guns with a 12.7mm (0.5in) Berezin UBS mounted ventrally between the wheel wells. The wing guns were removed, leaving the wings free for mounting air-to-air rockets, bombs or fuel. Against the faster Bf 109, the nimble I-16 was still competitive in the hands of a good pilot and many Soviet aces scored their first victories with it, but the German fighter was faster and better armed. By the end of 1943 virtually all surviving I-16s had been withdrawn to second-line roles and the last operational I-16 fighters, based at Petropavlovsk-Kamchatskiy in the far east of Russia, were exchanged for Bell Kingcobras as late as August 1945.

Given the tricky handling of the I-16, a trainer variant was designed early in its service life and built in large numbers as the UTI-1, 2 and 4. Exact production figures are unknown but may be as high as 3400. Total production of all I-16 variants was 10,292.

Polikarpov I-16 Type 24

Weight (Maximum take-off) 1882kg (4393lbs)

Dimensions Length: 6.1m (20ft 1in), Wingspan: 9m (29ft 6in), Height: 2.25m (7ft 5in)

Powerplant One 820kW (1100hp) Shvetsov M-63 9-cylinder supercharged air-cooled radial piston engine

Speed 489km/h (303mph)

Range 440km (270 miles)

Ceiling 10,800m (35,500ft)

Crew 1

Armament Four 7.62mm (0.3in) ShKAS machine guns; six unguided RS-82 rockets or up to 500kg (1100lbs) of bombs

Polikarpov I-16 Type 24

This winter camouflaged I-16 Type 24 was operating in the Lake Lagoda area, near Leningrad, during the winter of 1940–41 with 4 GvIAP of the Baltic Fleet.

Polikarpov I-16

'Red 28' was an anonymous I-16 with totally white upper surfaces that flew during the winter of 1941–42. Temporary white finishes were widely used in the first winters of the war.

Polikarpov I-185

Polikarpov's final fighter design was an outstanding aircraft destined through bad luck never to enter production. Despite this, prototypes of the I-185 were used in combat.

Polikarpov I-185

The I-185 was considered a marked improvement on its predecessor, the I-16, but never entered production.

In December 1938, test pilot Valeri Chkalov was killed in the prototype Polikarpov I-180, a development of the I-16. Unfortunately for Polikarpov, Chkalov was a national hero and a favourite of Stalin, and although the design was not found to be at fault and work was allowed to continue on the new fighter, Polikarpov's reputation never fully recovered.

A major redesign was implemented and during 1940 10 pre-series examples were built of the resultant I-180S. However, performance of this aircraft was deemed inadequate and a further redesign spawned the I-185. Intended for either the Tumansky M-90 or Shvetsov M-71 engine, which were both in the 1500kW (2011hp) class. Both engines were troubled but the M-71 eventually achieved sufficient reliability to power the first I-185 to fly in February 1942. The aircraft flew beautifully and performance was spectacular – the I-185 was faster than the Bf 109F at all altitudes and a speed of 685km/h (425mph) was ultimately recorded. The installed armament of three 20mm (0.78in) ShVAK cannon was impressive and like most Soviet fighters provision was made for a bombload of up to 500kg (1102lb) or eight RS-82 rockets.

Operational evaluation

In November 1942, three prototypes were sent to the front to be evaluated under operational conditions. The aircraft were not permitted to fly beyond the Soviet front line and only one firing pass was to be made on any enemy aircraft engaged. Results were unambiguously favourable, so plans to start production forthwith began and a 'production standard' aircraft was completed. Unfortunately both it and one of the prototypes were lost due to engine failure and the M-71 was beginning to be considered a dead end. Plans to produce the I-185 with the proven M-82 were abandoned as that engine was required for the inferior but adequate La-5 that was – significantly – already in full scale production. Despite its promise, the I-185 programme was formally cancelled in April 1943.

Polikarpov I-185 (M-71)

Weight (Maximum take-off) 3500kg (7716lbs)

Dimensions Length: 7.77m (25ft 6in), Wingspan: 9.8m (32ft 2in), Height: 2.5m (8ft 2in)

Powerplant One 1492kW (2000hp) Shvetsov M-71 18-cylinder air-cooled radial piston engine

Speed 630km/h (390mph)

Range 835km (519 miles)

Ceiling 11,000m (36,000ft)

Crew 1

Armament Three 20mm (0.78in) ShVAK cannon; eight RS-82 rockets, up to 500kg (1102lbs) of bombs

Lavochkin-Gorbunov-Gudkov LaGG-3

The overweight and underpowered LaGG-3 was an aircraft that was unloved by those who flew it and derided by its enemies.

Designed during 1938 by a team led by Semyon Lavochkin, Vladimir Gorbunov and Mikhail Gudkov, the 'LaGG' designation being a portmanteau of their surnames, the LaGG-3 answered a requirement for a new general purpose 'frontal' fighter with a best operating altitude of between 3000–4000m (9842–13,123ft) and intended as a somewhat overdue replacement for Polikarpov's ubiquitous I-16. Lavochkin and his associates proposed an aircraft constructed largely of wood to overcome the critical shortage of light alloys and steel then being experienced in the Soviet Union.

Composite construction

The design found ready favour due to both the limited demands it placed on strategic materials and performance estimates. Initially designated I-22, once work began on the new aircraft during the summer of 1939 it had been renamed I-301 (after Zavod 301, the factory where it was built). It featured a monocoque fuselage and wing spars both made of '*delta drevesina*,' a composite wooden material consisting of layers of birch strip bonded with Bakelite film, this material being both appreciably stronger than any untreated wood as well as fire resistant. Both wing and fuselage were covered with a stressed skin of Bakelite ply while the nose was a conventional structure of duralumin panels bolted to a steel tube framework. The engine was a Klimov M-105P and armament consisted of a single Volkov-Yartsev VYa-23V 23mm (0.9in) cannon firing through the airscrew hub and two 12.7mm (0.5in) Berezin UB machine guns mounted in the nose.

Overweight and underpowered

Work proceeded quickly, the aircraft being so urgently required that a production line for the first batch of 100 had been set up before the prototype, now re-designated LaGG-1, had even flown. Unfortunately, when the beautifully finished, hand-built first prototype did take to the air on 30 March 1940 it became clear that the aircraft was both overweight and underpowered. Although the speed performance was broadly in line with what was expected, acceleration was poor and the new fighter could be outclimbed by the I-16, the very aircraft it was supposed to replace. As if this wasn't bad enough the LaGG-1 demonstrated unpleasant handling

LaGG-1

Weight (Maximum take-off) 2968kg (6543lbs)
Dimensions Length: 8.81m (28ft 10in), Wingspan: 9.80m (32ft 2in), Height: 4.40m (14ft 5in)
Powerplant One 783kW (1050hp) Klimov M-105P V-12 liquid-cooled piston engine
Speed 605km/h (376mph)
Range 556km (345 miles)
Ceiling 9600m (31,490ft)
Crew 1
Armament One Volkov-Yartsev VYa-23V 23mm (0.9in) cannon and two 12.7mm (0.5in) Berezin UB machine guns

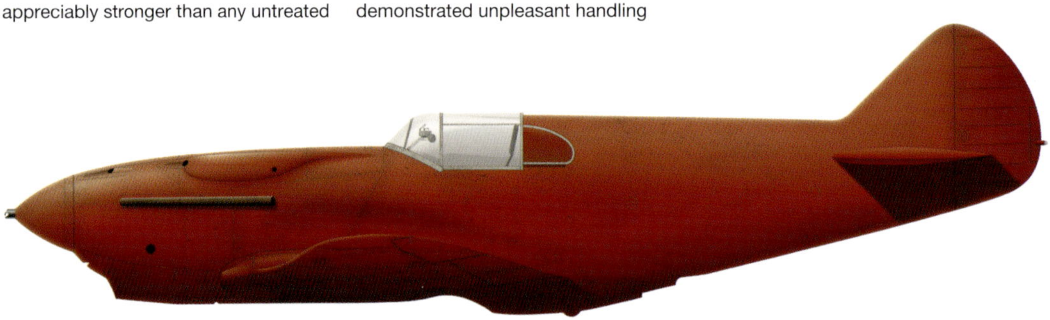

Lavochkin-Gorbunov-Gudkov LaGG-1

Noted for its fine finish, the I-301 was damaged after test pilot Nikishin was dazzled by the sun as he approached to land. Although repaired, it was written off shortly afterwards following a forced landing due to engine failure.

Lavochkin-Gorbunov-Gudkov LaGG-3

'White 57' was built in August 1942 and forced down by Finnish pilot Altto Tervo the following month. It was repaired and entered operational service with the Finns as LG-3.

characteristics and demanded a high degree of pilot skill. With no immediate prospect of more power, a crash programme of weight reduction was instigated while efforts were made to at least ameliorate the worst handling traits. At the same time it was deemed necessary that the aircraft have greater range so two additional 65 litre (14.3 gallons) wing fuel tanks were fitted and the aircraft was again re-designated as the LaGG-3.

The LaGG-3 in the form it began to roll off the production line was, at best, an immature combat aircraft. The wing benefitted from the addition of leading-edge slats to improve behaviour near the stall and handling had been improved by the addition of rather crude drop-shaped mass balance weights at the top and bottom of the rudder and a system of pendulum weights in the rear fuselage to balance the elevator. Later in production these would give way to dynamically balanced surfaces that arguably should have been fitted from the start. Even with these improvements the LaGG-3 was liable to develop a violent spin without warning in a steep banking turn, it had a tendency to nose-up on approach to land and would stall at the slightest provocation. The undercarriage was prone to collapse, the view from the cockpit was poor (a situation exacerbated by the opacity of Russian acrylic at the time), the hydraulics were unreliable, brakes seized, engines leaked oil, the gun firing mechanism often failed and initial build quality was bad. Compared to its great rival, the Yak-1, the LaGG-3 was marginally inferior in speed and climb rate but enjoyed a slightly better range. Although a much sturdier aircraft than the Yak, its poorer manoeuvrability and exacting handling characteristics did not endear it to service pilots.

Accelerated production

Despite its manifold shortcomings, production raced ahead and by 30

Lavochkin LaGG-3 66th series

Weight (Maximum take-off) 3190kg (7033lbs)

Dimensions Length: 8.81m (28ft 11in), Wingspan: 9.8m (32ft 2in), Height: 2.54m (8ft 4in)

Powerplant One 924kW (1239hp) Klimov M-105PF liquid-cooled V-12 engine

Speed 589km/h (366mph)

Range 1000km (620 miles)

Ceiling 9700m (31,800 ft)

Crew 1

Armament One 12.7mm (0.50in) Berezin BS machine gun, one 20mm (0.78in) ShVAK cannon; two 50kg (110lbs) bombs or six RS-82 or RS-132 rockets

Early-production LaGG-3s such as the one pictured here were considered basic aircraft, with crudely fashioned surfaces.

The same aircraft depicted as the three-view artwork opposite, this LaGG-3 of 524 IAP was shot down by Finnish anti-aircraft fire on 6 March 1942.

June 1941, 322 LaGG-3s had been completed and production would average 12.7 a day for the rest of the year.

In service the LaGG was unpopular, despite significant improvements to its handling and a concentrated effort to lighten the airframe, handling remained less than sparkling. Its only real advantage remained the immensely strong, fire-resistant airframe that could absorb remarkable amounts of punishment and remain airworthy. Even if a pilot overcame the LaGG's handling deficiencies it was tacitly admitted to be inferior to both the Bf 109 and Fw 190. Despite this, some pilots did manage to achieve a measure of success with the aircraft; a few even liked it. One such was ace Igor Kaberov whose affection for the Lavochkin fighter came about when he found the undercarriage could be retracted at the press of a button rather than having to be wound up by hand as on the I-16 he was used to. Top scorer is believed to have been Andrei Kulagin who shot down 26 aircraft with the type, unusually nearly all while flying the LaGG-3 at night.

Production was wound down and ceased in 1944, it having ceded front line fighter operations to its Yak rivals and the La-5 by this stage. The aircraft remained in service until the end of the war, with 6528 built in total.

Lavochkin-Gorbunov-Gudkov LaGG-3

Shown in temporary winter finish during late 1942, this LaGG-3 was flown by 15-victory ace Gerasim Grigoryev of 178 IAP over Moscow. The water-soluble white paint would wear off quickly, leading to a very careworn appearance.

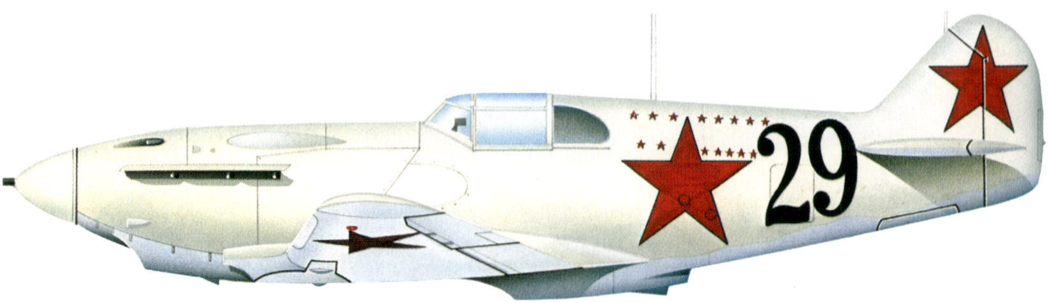

Lavochkin LaGG-3

Weight (Maximum take-off) 3190kg (7033lbs)

Dimensions Length: 8.81m (28ft 11in), Wingspan: 9.8m (32ft 2in), Height: 2.54m (8ft 4in)

Powerplant One 924kW (1239hp) Klimov M-105PF liquid-cooled V-12 engine

Speed 589km/h (366mph)

Range 1000km (620 miles)

Ceiling 9700m (31,800 ft)

Crew 1

Armament One 12.7mm (0.50in) Berezin BS machine gun, one 20mm (0.78in) ShVAK cannon; two 50kg (110lbs) bombs or six RS-82 or RS-132 rockets

Lavochkin-Gorbunov-Gudkov LaGG-3

Downed and repaired by the Finns, this aircraft, coded LG-1, scored the only confirmed kill by a LaGG-3 in Finnish service when it shot down a Soviet LaGG-3 on 16 February 1944.

Mikoyan-Gurevich MiG-1

Designed for the high altitude role, the first MiG fighter was produced in very small numbers before being supplanted by the MiG-3.

The MiG-1's origins lay in a January 1939 requirement for a fighter intended to achieve its maximum performance at altitudes above 6000m (19,685ft). Originally designed by Nikolai Polikarpov and designated I-200, the new aircraft took shape around the Mikulin AM-35A engine offering 1007kW (1350hp). With a dry weight of 830kg (1830lb), this engine compared unfavourably with contemporary rivals such as the Rolls-Royce Merlin at 624kg (1376lb) or Daimler-Benz DB60A of 608kg (1340lb). As a result the airframe was designed to be as small as possible and optimized for speed, this taking precedence over all other attributes.

Mikoyan Gurevich MiG-1

Weight (Maximum take-off) 3319kg (7317lbs)

Dimensions Length: 8.16m (26ft 9in), Wingspan: 10.2m (33ft 6in), Height: 2.62m (8ft 7in)

Powerplant One 1007kW (1350hp) Mikulin AM-35A V-12 liquid-cooled piston engine

Speed 657km/h (408mph)

Range 580km (360 miles)

Ceiling 12,000m (39,000ft)

Crew 1

Armament One 12.7mm (0.5in) Berezin UB machine gun, two 7.62mm (0.3in) ShKAS machine guns

Design shortcomings

Unfortunately for Polikarpov he lost favour with Stalin, and while on a tour of German aviation factories in November 1939 his engineering team was split up. The staff were distributed among other design bureaux and the nearly complete I-200 was handed over to a new bureau formed under Artem Mikoyan and Mikhail Gurevich. First flown on 5 April 1940, the small but heavy I-200 demonstrated excellent speed at high altitude, however in virtually every other regard the aircraft was found wanting. Longitudinal stability was poor, control response sluggish, it would fall easily into a vicious spin from which recovery was very difficult and landing characteristics were so bad it was recommended that only the most experienced pilots fly the aircraft.

Production improvements

Desperate for modern aircraft, an initial batch of 100 – now designated MiG-1 – was ordered in May 1940 despite the poor characteristics of the prototype. Production aircraft featured self-sealing fuel tanks and racks for a FAB-100 bomb under each wing. The troublesome side-opening canopy of the prototypes was replaced with a sliding canopy that could be jettisoned in flight if necessary. In total, 89 of the production aircraft were issued to fighter units by February 1941 and 55 were listed as operational before the German invasion in June. Details of the operational use of the MiG-1 remain obscure, although most were probably destroyed in the initial onslaught of Operation Barbarossa. However, one example was recorded on the VVS inventory as late as 1944.

Mikoyan-Gurevich MiG-1

Early-war fighters were painted with nitrocellulose lacquers that could be used on wood or fabric or on metal over an undercoat (usually yellow). Note the lighter shade of all green over the metal forward fuselage of this MiG-1 airframe operating with 31 IAP in June 1941.

Mikoyan-Gurevich MiG-3

Too specialized in concept for the combat conditions prevalent on the Eastern Front, the MiG-3 generally fought at a disadvantage. Despite this, several VVS aces scored their first kills with the type.

The MiG-1 had been found severely wanting as a combat aircraft and changes were introduced on the production line in an attempt to remedy the worst of its flaws. The most obvious external change was to the wing, which was given increased dihedral to improve lateral stability. The engine was moved forward 100mm (3.9in) that did little for the pilot's view forward but significantly improved longitudinal stability. Supercharger and radiator intakes were altered to improve airflow characteristics.

Increased fuel capacity

Internally, the poor range capability of the MiG-1 was improved with an additional fuel tank of 250 litre (55 gallon) capacity under the pilot's seat. Rearward view was improved somewhat with increased glazing behind the pilot and he now enjoyed the benefit of 8mm (0.3in) of armour

Mikoyan-Gurevich MiG-3

In temporary winter camouflage, 'Red 36' was operating with the 12 GvIAP in the defence of Moscow during the winter of 1941–42.

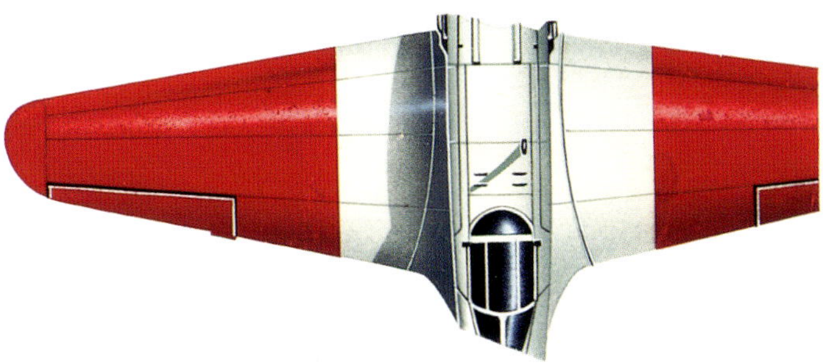

Mikoyan Gurevich MiG-3

Weight (Maximum take-off) 3355kg (7397lbs)

Dimensions Length: 8.25m (27ft 1in), Wingspan: 10.2m (33ft 6in), Height: 3.3m (10ft 10in)

Powerplant One 1007kW (1350hp) Mikulin AM-35A V-12 liquid-cooled piston engine

Speed 640km/h (400mph)

Range 820km (510 miles)

Ceiling 12,000m (39,000ft)

Crew 1

Armament One 12.7mm (0.5in) Berezin UBS machine gun, two 7.62mm (0.3in) ShKAS machine guns

Mikoyan-Gurevich MiG-3

With crudely applied extra camouflage and discarded cockpit canopy, this MiG-3 was operating in the defence of Leningrad during the summer of 1941.

protecting his back. Installed armament remained the same trio of one 12.7mm (0.5in) UBS machine gun and a pair of 7.62mm (0.3in) ShKAS, all mounted in the nose to provide a combined firepower that was generally considered inadequate by late 1940. Indeed, to improve the armament 821 aircraft were later built with two Berezin 12.7mm UKB machine guns in pods under each wing but the extra weight and drag of these weapons impaired performance and they were soon discarded.

Build quality problems

First flown in October 1940, the MiG-3 had gained 250kg (551lb) in loaded weight over the MiG-1 and though the

Pilots disperse after listening to a speech on combat tactics given by Aleksander Pokryshkin, the Soviet Union's third highest scoring ace (with 59 kills by the end of the war, 12 scored while flying the MiG-3) whose first taste of action came on the opening day of Operation Barbarossa. The 'parade ground' line-up of MiG-3s in the background (which also includes Lend-Lease Curtiss P-40s) was typical; the VVS lost many fighters in the opening bombing raids of Barbarossa because they failed to disperse their aircraft to minimize potential losses in such raids.

Mikoyan Gurevich MiG-3
Weight (Maximum take-off) 3355kg (7397lbs)
Dimensions Length: 8.25m (27ft 1in), Wingspan: 10.2m (33ft 6in), Height: 3.3m (10ft 10in)
Powerplant One 1007kW (1350hp) Mikulin AM-35A V-12 liquid-cooled piston engine
Speed 640km/h (400mph)
Range 820km (510 miles)
Ceiling 12,000m (39,000ft)
Crew 1
Armament One 12.7mm (0.5in) Berezin UBS machine gun, two 7.62mm (0.3in) ShKAS machine guns

MiG-3 was faster at all altitudes, its climb rate had decreased and it now required a longer field length to get airborne. Performance also suffered on early production machines due to poor build quality, a situation dealt with by the authorities in a brutally simple fashion by removing a number of senior managers on trumped-up charges of sabotage. Range capability was found to be lower than the 1000km (621 miles) demanded by the specification but Mikoyan disagreed with the findings of the Test Institute and staged his own demonstration by flying two standard production aircraft from Moscow to Leningrad over routes of 970–1015km (602–631 miles), not taking into account ascent and descent.

The range of the MiG-3 was recalculated and the failure of the Air Force Scientific Test Institute to correctly determine this figure was summarily dealt with by executing its head, Major General A.I. Filin, a test pilot of considerable experience.

Mikoyan-Gurevich MiG-3

'Red 02' was another 12 GvIAP MiG-3. Red panels on the wing surfaces were intended to assist search aircraft in the event the aircraft was forced down.

Mikoyan-Gurevich MiG-3

Operating with the 7 IAP, 'White 04' was flown by Sergey Polyakov in the defence of Leningrad in the summer of 1941. An I-16 veteran of the Spanish Civil War, during 1941 Polyakov would transfer to an Il-2 *Shturmovik*.

Each of these MiG-3s carries RS-82 rockets under its wing. Up to eight could be carried, as an alternative to a pair of 100-kg (220-lb) FAB-100 bombs in the fighter-bomber role.

With the pressing need for modern fighters overruling concerns about the more problematic aspects of the new MiG, production was initiated in February 1941 at Zavod 1. A second factory tooled up during March, and by April a weekly total of 75 aircraft was being produced.

For organizing such a rapid initiation of production both Artem Mikoyan and Mikhail Gurevich received the State Stalin Prize. Unfortunately for front line squadrons, production tempo far exceeded the number of pilots trained to fly the MiG-3 – for example, on 1 June 1941, the VVS had over 1000 MiG-3s on strength but fewer than 500 trained pilots.

Pilot skill deficit

The arrival of the new fighter was not received with universal delight. Although the worst aspects of the MiG-1 had been ameliorated, the MiG-3 was still an aircraft with some difficult characteristics and demanded a high level of piloting skill, arguably more than that possessed by the average VVS fighter pilot. Even before hostilities began, losses were commonplace. For example, in April 1941 three MiG-3s dispatched to intercept a German reconnaissance aircraft flying at 9000m (29,530ft) fell into irrecoverable spins with all three pilots bailing out, one of whom was killed.

Once committed to actual combat the shortcomings of the MiG became starkly obvious. Although faster than its principal opponent, the Messerschmitt Bf 109F, at the high altitudes that it had been designed for, at the medium to low altitudes where virtually all combat on the Eastern Front took place the MiG-3 enjoyed no speed advantage. The MiG-3 also possessed a loaded weight some 220kg (485lb) greater than the Messerschmitt and was less manoeuvrable in the horizontal plane. Despite this It was possible to fly the

MiG effectively in combat and several pilots scored five or more kills with the type. One of these was Aleksander Pokryshkin, the second highest scoring Allied fighter pilot, who destroyed the first five of his officially credited 59 kills with the type.

Final MiG-3s

With two other more versatile fighters in production and faced with an acute shortage of attack aircraft, the decision was made in December 1941 to terminate production. This decision was also influenced by the fact that the MiG-3's Mikulin AM-35A was closely related to the Il-2's AM-38 and the factory producing the motor could switch production with minimal delay. The last few MiG-3s were assembled at Zavod 1 from surplus components in April 1942, the factory having already switched to *Shturmovik* production.

During 1942, surviving MiG-3s were increasingly assigned away from the front, defending cities and other strategic targets. Here they enjoyed more success than as a 'frontal' fighter and some enjoyed quite lengthy service – the last operational MiG-3s were only withdrawn as the war in Europe drew to an end.

Mikoyan Gurevich MiG-3
Weight (Maximum take-off) 3355kg (7397lbs)
Dimensions Length: 8.25m (27ft 1in), Wingspan: 10.2m (33ft 6in), Height: 3.3m (10ft 10in)
Powerplant One 1007kW (1350hp) Mikulin AM-35A V-12 liquid-cooled piston engine
Speed 640km/h (400mph)
Range 820km (510 miles)
Ceiling 12,000m (39,000ft)
Crew 1
Armament One 12.7mm (0.5in) Berezin UBS machine gun, two 7.62mm (0.3in) ShKAS machine guns; two 100kg (220lbs) bombs or up to eight RS-82 or rockets

Mikoyan-Gurevich MiG-3

One of the Soviet Union's most successful pilots, Alexander Pokryshkin scored his first 20 victories with the MiG-3. Pokryshkin flew 'White 5' with the 16 GvIAP during 1942.

Mikoyan-Gurevich MiG-3

Still in frontline service in 1944, this MiG-3 was operating with 7 IAP of the Black Sea Fleet, in concert with P-40s in the Kuban bridgehead area.

Petlyakov Pe-3

The Pe-2 bomber design started life as a fighter but development came full circle with the Pe-3 fighter derivative. Only 360 were built but they saw extensive service.

One of the first Pe-2s had been reworked as a Pe-3 fighter variant in the spring of 1941, but maximum priority at the time was for bombers, so production was not ordered. This situation changed markedly when German night raids began against Moscow in July and a crash programme was initiated to field a suitable night fighter. When the Pe-3 first entered production it featured the bare minimum of changes required to allow it to operate in its new role.

Design variants

Distinguishable from its early Pe-2 brethren by the reduced nose glazing, the Pe-3 mounted two 12.7mm (0.5in) UB and one 7.62mm (0.3in) ShKAS machine guns in the nose. Slats to improve manoeuvrability took the place of the Pe-2's wing-mounted dive brakes and the bomb bay was used for fuel.

Although the ventral gun position was replaced by more fuel tankage, the dorsal 7.62mm (0.3in) was retained, along with the bomb bays at the rear of the engine nacelles. A fixed ShKAS was fitted in the tail cone firing directly rearwards, but this was of limited value and deleted.

Rushed production

The urgency attached to Pe-3 production may be judged by the fact that the initial conversion was ordered to be completed in a mere four days and production of the new fighter began even before proper drawings had been prepared. Instead workers fabricated parts based on sketches and adjusting them to fit by hand.

Petlyakov Pe-3
Weight (Maximum take-off) 8000kg (17,637lbs)
Dimensions Length: 12.66m (41ft 6in), Wingspan: 17.13m (56ft 2in), Height: 3.5m (11ft 6in)
Powerplant Two 820kW (1100hp) Klimov M-105RA liquid-cooled V-12 engines
Speed 530km/h (330mph)
Range 2150km (1336 miles)
Ceiling 9100m (29,900ft)
Crew 2
Armament One 20mm (0.78in) ShVAK cannon in the nose, two 12.7mm (0.5in) UBK machine guns in the fuselage, one 12.7mm (0.5in) UBT machine gun in dorsal turret, one 7.62mm (0.3in) ShKAS machine gun fixed firing rearward in tail cone (later deleted); up to 700kg (1543lbs) bombload

First deliveries occurred within a month of the initiation of the Pe-3 programme and combat experience revealed a need for armour protection and increased firepower.

Pe-3bis

These concerns were addressed in the Pe-3bis, which added a ShVAK 20mm (0.78in) cannon in the nose, along with armour and relocated weaponry to avoid blinding the pilot when fired at night. In service the Pe-3 was worked hard as a night fighter and ground attack aircraft. Ultimately its high speed and range saw it increasingly used as a reconnaissance platform, a role it performed until the end of hostilities.

This photographs clearly shows the 20mm (0.78in) ShVAK cannon in the nose, as well as the dorsal 12.7mm (0.5in) machine gun. Built in small numbers, the Pe-3 proved to be a durable and adaptable heavy fighter, employed as both a night interceptor and ground-attack aircraft.

Petlyakov Pe-3

With its field-applied temporary white camouflage, this Pe-3 operated with 6 IAK of the Moscow PVO during the winter of 1941–42. This aircraft retains the rear-firing fixed ShKAS machine gun in the tail cone.

Yakovlev fighters

When the I-26 first flew on 13 January 1940, few could have predicted that this thoroughly conventional aircraft heralded the start of the greatest fighter production programme in history.

Built to the same 1938 requirement as the Lavochkin-Gorbunov-Gudkov LaGG-3 for a general purpose tactical fighter, the I-26 as it was initially known was the first fighter design of the young designer Aleksander Yakovlev. In contrast to its competitor, the I-26 was light and easy to handle. Its performance earned Yakovlev the Order of Lenin, a 100,000 rouble prize and a car.

Ordered into production as the Yak-1 a mere month or so after the prototype's first flight, the aircraft was small, simple and of mixed wood and metal construction to minimize demand on strategic light alloy and steel supplies. It turned out to be ideally suited to the operational conditions it would encounter during the Great Patriotic War. To achieve a decent performance on the relatively modest 940kW (1260hp) at 700m (2296ft) delivered by its Klimov M-105, the Yak-1 adopted diminutive proportions – in fact it was the lightest combat aircraft of its generation. Installed armament was similarly minimal, consisting of just one 20mm (0.78in) ShVAK cannon firing through the propeller hub augmented by two 7.62mm (0.3in) ShKAS machine guns in the upper nose decking.

Teething problems

Unfortunately, although admittedly beset with fewer problems than its LaGG-3 contemporary, the Yak-1 had been ordered into production a mere two months after the prototype first

Yak-1

'Yellow 44' was flown by Lydia Litvyak of 73 GvIAP, the most successful female fighter pilot in history. It is not definitely known how many aircraft she shot down, estimates ranging from 5 to 12, all achieved while flying the Yak-1.

A Yak-1 fighter leaves the assembly shop. Early Yak-1s were outclassed by their German opponents, but the type soon developed a reputation for reliability and became popular with Russian pilots.

Yak-1

Mikhail Baranov was the leading Soviet ace of 1942, flying this Yak-1 over Stalingrad marked with 27 victory stars and the slogan '*Smert Fascistam*' ('Death to Fascists') rather crudely applied to the rear fuselage.

Yak-1

A Yak-1 of an unknown unit operating over central Russia during the winter of 1941–42. With the spring thaw, the white paint would be washed off and the aircraft returned to its standard black-and-green scheme.

flew and numerous problems had to be overcome as the aircraft began to roll off the production line. As many as 7460 changes were implemented on the assembly line in the first four months alone. As a result aircraft were essentially hand-finished with components adapted to fit individual airframes. Yak-1s ended up with bizarre problems such as different length landing gear legs, parts that were not interchangeable between aircraft and no two early Yaks could be relied upon to have the same performance. Teething problems remained: landing gear collapsed, oil overheated, plywood skins detached and fuel lines ruptured. Nonetheless, the Yak-1 was at heart, a pilot's aeroplane and demonstrably the best of the domestic fighter types that would be called on to face the German juggernaut in the summer of 1941. Western military districts had received 105 Yak-1s and only the 20th IAP near Kiev had partially completed conversion training. As a result the Yak-1's contribution to repulsing the invaders was limited, 20 IAP losing all but six of its 63 aircraft in the first month of hostilities.

As the Yak-1 became a greater part of the Soviet fighter component, strenuous efforts were undertaken to cure the lingering problems that still afflicted the aircraft and by 1942 the Yak-1 was a reliable aircraft and very popular with pilots. Modifications were introduced onto the production line – from the 1000th aircraft onward radios began to be fitted, initially to every

Yakovlev Yak-1 (early production)

Weight (Maximum take-off) 2844kg (6269lbs)

Dimensions Length: 8.48m (27ft 10in), Wingspan: 10m (32ft 10in), Height: 2.64m (8ft 8in)

Powerplant One 783kW (1050hp) Klimov M-105PA V-12 liquid-cooled piston engine

Speed 560km/h (348mph)

Range 700km (430 miles)

Ceiling 9900m (32,500ft)

Crew 1

Armament One 20mm (0.8in) ShVAK and two 7.62mm (0.3in) ShKAS machine guns

10th airframe, the twin ShKAS 7.62mm guns were replaced with a single 12.7mm (.50in) Berezin UBS and from October 1942 a cutdown rear fuselage and bubble canopy were fitted. Arguably the most important change was the introduction of the M-105PF 'boosted' engine that, coupled with a lightened airframe, produced a marked improvement in performance.

Back in 1939 Yakovlev had designed a two-seat derivative of his Yak-1, the I-27, intended primarily as an advanced trainer although able to take on fast courier and light transport duties when required. First flown in the late summer of 1940 and now known as the UTI-26, the new aircraft differed from the YAK-1 solely in possessing a marginally bigger wing of 25cm (9.8in) greater span, slightly larger tail surfaces and a second cockpit with full flight controls. The wing was placed further back and the radiator was moved forward to maintain the centre of gravity, which also happened to improve the stability of the aircraft. The undercarriage was simplified and strengthened and armament was reduced to a single ShKAS in the nose decking.

Yak-7UTI

Such was the importance ascribed to the UTI-26 that it was ordered into production before the conclusion of its State Acceptance trials. The first production machines, now with the definitive Yak-7UTI designation, appeared in May 1941. In August Engineer Sinelshchikov at Zavod 301 decided that a single-seat derivative of the Yak-7UTI might prove effective

Yakovlev Yak-7 (early production)
Weight (Maximum take-off) 2935kg (6471lbs)
Dimensions Length: 8.48m (27ft 10in), Wingspan: 10m (32ft 10in), Height: 2.75m (9ft)
Powerplant One 780kW (1050hp) Klimov M-105PA V-12 liquid-cooled piston engine
Speed 495km/h (308mph)
Range 643km (400miles)
Ceiling 9500m (31,200ft)
Crew 1
Armament One 20mm (0.78in) ShVAK cannon and two 12.7mm (0.5in) Berezin UBS machine guns

Yak-7B

Pictured as it appeared in the autumn of 1942, this bubble canopied late-production Yak-7B served with 42 IAP. Note the 'kill' markings on the fuselage scroll motif – it was highly unusual to record victories with swastikas in the VVS.

Yak-7

The hump-backed profile of the Yak-7 belie its trainer origins. This example, still wearing the remains of its winter camouflage, was serving with 29 GvIAP at Leningrad during the spring of 1943.

Yak-9T-37

Serving with the French Normandie-Niemen Regiment, 'White 60' was a Yak-9T-37 variant equipped with a 37mm (1.45in) cannon firing through the spinner. It was flown by René Challe in June 1944.

and altered a standard Yak-7UTI back into a fighter. Armament was restored to one 20mm (0.78in) ShVAK and two 12.7mm (0.5in) Berezin UBS guns in the nose though the 'new' fighter gained racks for six rockets under each wing. The rear cockpit canopy remained unchanged so externally the aircraft looked exactly the same as its trainer brethren. In early Yak-7 production the rear cockpit was faired over, endowing it with a distinctive hump-backed profile. Flight testing revealed the single seater was superior to the Yak-1 and retained the useful ability to carry a passenger behind the cockpit. Serial production was simple as the aircraft was a modification of an existing type and the first Yak-7s were in action towards the end of the Battle of Moscow in December 1941.

Both the Yak-1 and Yak-7 were produced simultaneously, the Yak-7 receiving the cutdown rear fuselage at the same time as the Yak-1. Although most of the Yak-7s built were standard fighters, a small batch of 22 Yak-7-37s with a 37mm (1.4in) Shpital'ny-Komaritsky cannon firing through the propeller hub was built and combat tested during August 1942, but no further production resulted. Other experimental versions were the Yak-7PD high-altitude prototype with an 11,300m (37,073ft) ceiling and the YAk-7 M-82, which was a radically reworked Yak-7 fitted with a Shvetsov M-82 radial engine. By 1943 major structural changes, particularly to the wing, warranted a designation change to Yak-9, by which time 6399 of both single and twin seat Yak-7s had been built.

Yak-9

Two distinct 'heavyweight' and 'lightweight' lines of development of the basic Yak fighter emerged and the various Yak-9 models represented the culmination of the development of the more versatile 'heavy' version derived from the Yak-7. The terms were of course relative – compared to the fighters of most nations the Yak-9 was still a featherweight aircraft.

First appearing on the front during late November 1942, the Yak-9 differed from the Yak-7 primarily in its increased fuel capacity in the wings that were now primarily constructed of light alloy. One of the 12.7mm (0.5in) UBS guns was deleted, and when loaded the initial production Yak-9 was appreciably lighter than its Yak-7 predecessor. Despite the greater fuel capacity, yet more range was demanded and the

Yak-9T-37

Weight (Maximum take-off) 3025kg (6668lbs)
Dimensions Length: 8.65m (28ft 4in), Wingspan: 9.74m (31ft 11in), Height: 2.75m (9ft)
Powerplant One 1014kW (1360hp) Klimov M-105PF-3 V-12 liquid-cooled piston engine
Speed 597km/h (371mph)
Range 1400km (870 miles)
Ceiling 10,000m (32,750ft)
Crew 1
Armament One 37mm (1.45in) Nudelman-Suranov NS-37 cannon, one 12.7mm (0.5in) Berezin UBS machine gun

Yak-3

A Yak-3 of the 303rd Fighter Aviation Division, First Soviet Air Army, on the 3rd Ukrainian Front in 1944. The pilot of this aircraft was 303rd commander Georgii Nefedovich, who eventually accumulated 23 kills.

Yak-3

Weight (Maximum take-off) 2697kg (5946lbs)

Dimensions Length: 8.5m (27ft 11in), Wingspan: 9.2m (30ft 2in), Height: 2.64m (8ft 8in)

Powerplant One 960kW (1290hp) Klimov VK-105PF2 V-12 liquid-cooled piston engine

Speed 646km/h (401mph)

Range 550km (340 miles)

Ceiling 10,400m (34,100ft)

Crew 1

Armament One 20mm (0.78in) ShVAK cannon and two 12.7mm (0.5in) Berezin UBS machine guns

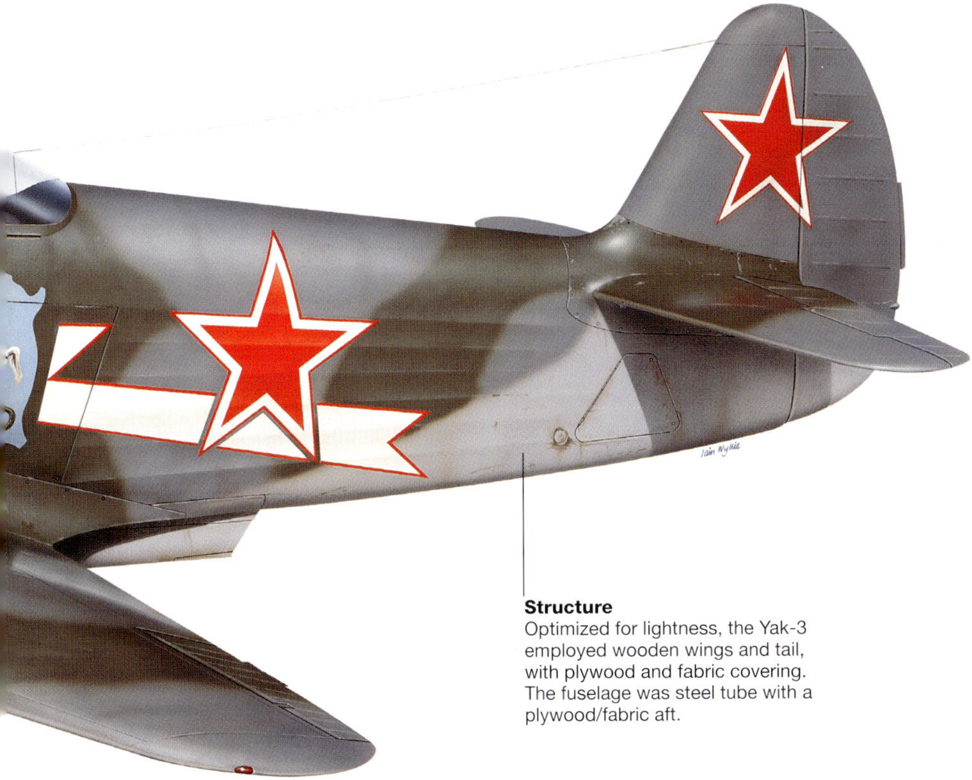

Structure
Optimized for lightness, the Yak-3 employed wooden wings and tail, with plywood and fabric covering. The fuselage was steel tube with a plywood/fabric aft.

Yak-9D for *dal'ny* 'long' (range) was the result. An internal capacity of 650 litres (142.9 gallons) increased range to 1400km (870 miles), accompanied with an increase in loaded weight that reduced climb rate although top speed was little affected. Later the Yak-9DD *dal'ny deistviye* 'ultra long' increased this to a maximum of 2300km (1429 miles), this variant being employed to escort USAAF B-17s and B-24s on shuttle bombing raids in the summer of 1944.

A separate line of development sought to optimize the aircraft for the fighter bomber role by incorporating an internal bomb bay into the design, the Yak-9B *bombardirovshchik* crammed four 100kg (220lb) bombs nose-up directly behind the cockpit.

The Yak-9B was intended for pinpoint attacks against heavily defended targets. Service use was somewhat disappointing due to difficulties in bomb aiming and poor handling when fully loaded, this being the heaviest Yak-9 variant to see service. The contemporary Yak-9T *tyazhely* 'heavy' mounted a Nudelman-Saranov NS-37 37mm (1.45in) cannon with 30 rounds firing through the propeller hub.

In total 2748 of this variant were produced and they were used to devastating effect against tanks and shipping. An even heavier armed version with a hub-mounted 45mm (1.7in) NS-45 cannon with distinctive muzzle brake was developed as a bomber destroyer but saw relatively little use due to recoil problems and a decline in German air activity by the time it entered service.

Yak-3 and Yak-1M

The last Yak fighter to be developed during the war was the Yak-3, which had started life as the Yak-1M for *Moskit* ('Mosquito'). This was a lightened version, intended to maximize air-to-air combat performance, with wings of reduced span and aerodynamic improvements, most obviously the relocation of the oil cooler from under the nose to the wing roots.

First flown in February 1943, the Yak-1M was tested against captured examples of the Fw-190A-4 and Bf-109G-2 and was found to possess better performance than either at low and medium altitude. Test pilots were unanimous in their praise and the Yak-1M was rushed into production as the second iteration of the Yak-3 designation, the first Yak-3 having been a metal-winged Yak-1 derivative known as the I-30 that was abandoned at the outbreak of war.

YAK-9 POST-WAR SERVICE

Later development focused on the Yak-9U, featuring the new Klimov VK-107 that offered 1300kW (1743hp) on take-off and promised spectacular performance, the prototype attaining 700km/h (435mph) at 5600m (18,372ft). Teething issues with this engine meant that only a handful of these exceptional fighters made it into service before the end of hostilities. The Yak-9U would be produced until 1948, receiving the NATO reporting name Frank, and it formed the basis of several post-war air forces, not least that of North Korea where it would once again see combat.

Yak-3

Weight (Maximum take-off) 2697kg (5946lbs)
Dimensions Length: 8.5m (27ft 11in), Wingspan: 9.2m (30ft 2in), Height: 2.64m (8ft 8in)
Powerplant One 960kW (1290hp) Klimov VK-105PF2 V-12 liquid-cooled piston engine
Speed 646km/h (401mph)
Range 550km (340 miles)
Ceiling 10,400m (34,100ft)
Crew 1
Armament One 20mm (0.78in) ShVAK cannon and two 12.7mm (0.5in) Berezin UBS machine guns

Yak-3

'Blue 24' was based in East Prussia by April 1945 with the 7 GvIAP of the 303 IAD.

Yak-3

Featuring a stylized winged sword on the nose, Yak-3 'White 100' was flying with 492 IAP as the unit entered Germany in early 1945.

Initial deliveries were slow, despite the basic aircraft being very closely related to the Yak-1 it supplanted on the assembly line. This was because each Yak-3 required 2.5 times as many man-hours to complete, mostly due to the care required in producing a high-quality surface finish. Even then, production Yak-3s could not attain the same performance as the Yak-1M prototypes due to poorer build quality. By the start of 1945 considerable numbers of the new fighter were available – in January the VVS reported 735 Yak-3s on strength.

The Yak-3 delighted its pilots, and for the first time the German fighter pilots admitted a Soviet fighter was superior to their own, a 1944 *Luftwaffe* general directive famously advised pilots to 'avoid combat below 5000m (16,400ft) with Yakovlev fighters lacking an oil cooler under the nose.'

More firepower

By 1945 much effort was being expended on improving the armament, culminating in the Yak-3P that featured three Berezin B-20 20mm (0.78in) cannon – two in the nose decking and one firing through the airscrew hub. Although too late for an active combat role, 596 Yak-3Ps were built from April 1945 onwards.

The most radical development was the Yak-3R that featured a RD-1khZ rocket in the tail, which boosted speed to a recorded maximum of 782km/h (486mph). Production of this, the fastest of the Yaks, was not undertaken due to the promise offered by turbojet development. Total production of all piston engine Yak fighters amounted to 36,737, of which 2190 were built in the post-war period.

Normandie-Niemen Regiment Yak-3s are seen shortly after the unit re-equipped with the new fighter type in July 1944. Yak-1s and Yak-9s had earlier equipped the Free-French squadrons.

Yak-1M

From October 1942, Yak-1s were built to an improved standard with bubble canopy, lowered rear fuselage and many internal changes. This example, presented by the 'Collective Workers of Shatovskovo Village Soviet' was flown by Hero of the Soviet Union, 35-kill ace, Aleksey Reshetov.

Yak-1M

Another high-scoring Yak-1 pilot, and twice Hero of the Soviet Union, was Sergey Lugansky, who had scored 32 of a total 37 victories, as recorded in the laurel wreath. Lugansky flew this example presented by the 'Youth and Young Communists of Alma Ata' during 1944.

Yakovlev Yak-1M (late production)

Weight (Maximum take-off) 2884kg (6358lbs)
Dimensions Length: 8.48m (27ft 10in), Wingspan: 10m (32ft 10in), Height: 2.64m (8ft 8in)
Powerplant One 940kW (1260hp) Klimov M-105PF V-12 liquid-cooled piston engine
Speed 592km/h (368mph)
Range 700km (430 miles)
Ceiling 10,050m (32,970ft)
Crew 1
Armament One 20mm (0.8in) ShVAK cannon and one 12.7mm (0.5in) Berezin UBS machine gun

Yak-1M

The tricolour spinner and French roundel beneath the windscreen identify this late Yak-1 as operating with the French-manned Normandie-Niemen Regiment. This Yak served with the première escadrille of the regiment in 1943.

Lavochkin La-5

A switch to the brilliant M-82 radial engine reversed the fortunes of Semyon Lavochkin, the La-5 allowing pilots to combat German fighters on broadly level terms for the first time.

Lavochkin La-5FN

Flown by Nikolai Pushkin of 2 GvIAP in September 1943, La-5FN 'White 52' was a presentation aircraft presented to the VVS by the Mongolian People's Republic.

While not exactly in disgrace, by 1942 the ongoing poor performance of Lavochkin's LaGG-3 had led to the reduction of the designer's team to a bare nucleus. They were based at a remote Georgian factory to which LaGG-3 production had been transferred from factories at Sevastopol and Taganrog, which were now in German hands.

Air-cooled engine

The adaptation of existing inline-engined single seat fighter aircraft had been pursued as an official directive from the summer of 1941, the primary purpose being to safeguard against any potential shortfall in inline engines. In due course the Yak-7, Mig-3 and LaGG-3 would all be re-engined but the benefit afforded to the Lavochkin aircraft was nothing short of extraordinary. The new engine was the Shvetsov M-82 offering 1230kW (1649hp) at take-off. The M-82 was 250kg (551lb) heavier than the Klimov engine of the LaGG-3 and adapting the slender fuselage for the 777mm (30.5in) width of the radial was a not a simple process, made even more complicated by the differing thrust lines of the two engines and centre of gravity changes resulting from the deletion of the ventral radiator that was not required for an air-cooled engine. This last aspect allowed the fitment of only the lightest permissible armament of two ShVAK 20mm (0.78in) cannon in the nose.

The radial engine of the converted fighter was extremely closely cowled, sufficient engine cooling being achieved through the use of fans mounted in front and behind the engine. Adjustable louvres controlled the flow of cooling air into the engine and the exhaust pipes were arranged in a vertical cluster on either side of the fuselage and delivered some thrust augmentation.

Semyon Lavochkin reinstated

Completed in December 1941 approximately two months after design

Lavochkin La-5FN

Weight (Maximum take-off) 3402kg (7500lbs)
Dimensions Length: 8.67m (28ft 5in), Wingspan: 9.8m (32ft 2in), Height: 2.54m (8ft 4in)
Powerplant One 1380kW (1850hp) Shvetsov M-82FN 14-cylinder air-cooled radial engine
Speed 648km/h (403mph)
Range 765km (475 miles)
Ceiling 11,000m (36,000ft)
Crew 1
Armament Two 20mm (0.78in) ShVAK cannon; two bombs up to 100kg (220lbs) each

work began, the conversion had to wait until March before flight testing could begin due to bad weather and problems encountered with cooling the M-82 engine.

First flight data revealed improved climb and speed performance and after some control issues were dealt with, the aircraft was submitted for accelerated State Acceptance trials. The enthusiasm of the test pilots for the new aircraft resulted in a priority directive issued in July that all existing LaGG-3 airframes in production should be converted to the M-82 immediately, simultaneously Semyon Lavochkin was restored to his previous position in control of Zavod 21 at Gor'kiy and ordered to establish production of the La-5, as it had been designated, as expeditiously as possible.

La-5 design issues

The first 10 conversions went to a Trials unit that was expected to evaluate the La-5 while flying it on operations. These first La-5s, like much contemporary Soviet equipment, exhibited poor build quality and performance was significantly worse than the prototype. Even worse, two of the first production machines fatally crashed due to losing a wing. These issues were promptly resolved and the La-5 was committed to action from September 1942 over Stalingrad.

The fighter now proved very combat worthy – it retained the immensely sturdy structure of the LaGG-3, but combined it with an excellent handling airframe with exceptional powers of manoeuvre. The aircraft was brilliant but not perfect: armament was adequate at best and the La-5 displayed a propensity to bounce alarmingly on landing that was never fully eradicated. Pilots would joke that, 'the only Lavochkin that doesn't bounce is a Focke-Wulf'.

Fighter ace

As production got into its stride the La-5 received a cut down rear fuselage to improve visibility and from late 1942 the M-82F *forsirovanny* 'boosted' offered a useful increase in power, followed in March by the even more potent M-82FN *forsirovanny neprosredstvenno* 'directly boosted' engine that featured fuel injection and delivered 1380kW (1850hp) at take-off.

The definitive La-5FN appeared in numbers over the front during March 1943, and German aviators reported that Soviet pilots in La-5s in particular were being flown with more aggressiveness than had previously been displayed.

It was during this time that the career of Ivan Kozhedub, the most successful Allied fighter pilot, began its meteoric rise and he would go on to score most of his victories with the La-5FN. La-5FNs also served with the 1st Czechoslovak Fighter Regiment, most dramatically in support of the Slovak National Uprising during September and October 1944.

These Lavochkin La-5FNs, operated by the 1st Czech Fighter Regiment, were photographed on 11 September 1944 immediately prior to a flight from Proskurov in Ukraine to Stubno.

Lavochkin La-5F

Relatively few La-5s were finished in the standard early war black-and-green fighter scheme. 'White 90' was serving with 523 IAP during the winter of 1943–44.

Lavochkin La-5FN

Vitaly Popkov of 2 GvIAP scored the majority of his 40 victories with the La-5FN. 'Yellow 01' was his personal aircraft in the summer and autumn of 1944 while operating over the Ukrainian Front.

Lavochkin La-5FN

'White 69' of 1 Czechoslovak IAP was flown by former 310 Squadron RAF Spitfire pilot Ladislav Valousek over Ukraine during September 1944. The diamond marking on the engine cowl is a stylized representation of the cyrillic letters 'FN', denoting the engine variant.

Lavochkin La-5FN

This Lavochkin La-5FN was flown by Kapitan Petr Yakovlevich Likholetov of the 159th Fighter Aviation Regiment. Likholetov's final score was 30 enemy aircraft destroyed. Seriously injured in a car crash at the end of 1944, he died of his injuries on 13 July 1945.

Lavochkin La-5FN

Weight (Maximum take-off) 3402kg (7500lbs)
Dimensions Length: 8.67m (28ft 5in), Wingspan: 9.8m (32ft 2in), Height: 2.54m (8ft 4in)
Powerplant One 1380kW (1850hp) Shvetsov M-82FN 14-cylinder air-cooled radial engine
Speed 648km/h (403mph)
Range 765km (475 miles)
Ceiling 11,000m (36,000ft)
Crew 1
Armament Two 20mm (0.78in) ShVAK cannon; two bombs up to 100kg (220lbs) each

Armament
The La-5 was equipped with two 20mm (0.78in) ShVAK cannon in the engine cowling with 200 rounds per gun.

Engine
A powerful Shvetsov M-82FN engine outputted 1380kW (1850hp), offering a maximum speed of 648km/h (403mph).

Canopy
The canopy of the La-5 offered excellent all-round vision, but was difficult to open at high speeds. Consequently, some pilots flew with their canopies open.

Lavochkin La-7

An exceptional combat aircraft, the La-7 maintained the hard-won Soviet air superiority over the Eastern Front.

Lavochkin La-7

Sergei Dolgushin flew this La-7 when commander of 156 IAP over Germany in the spring of 1945. Note victory stars and 'Hero of the Soviet Union' medal on the cowling. Dolgushin survived the war and flew with the VVS until 1970!

The La-7 was not a new design but represented a relatively modest aerodynamic refinement of the proven La-5 airframe. It was essentially an interim model before the completely new all-metal La-9 could be brought into service. In the event the latter aircraft first flew during 1946 and the La-7 remained the ultimate Lavochkin design to see action during World War II, as well as being the last fighter of predominantly wooden construction to see widespread service.

Aerodynamic efficiency

Developed from the La-5FN, the most obvious external difference to distinguish the La-7 from its progenitor was the removal of the prominent supercharger air intake above the cowling to the port wing root to improve pilot view. To refine the aerodynamic efficiency of the engine cowling the oil cooler under the nose was repositioned under the fuselage, level with the wing trailing edge, and the starter dog was removed from the airscrew spinner. The cowling itself consisted of two hinged panels aft of the front ring that swung upwards to reveal the engine and provided excellent access for maintenance.

First taking to the skies in November 1943, and known at this point as the La-5 mod.1944, the aircraft demonstrated a top speed of 680km/h (422mph) at 3000m (9842ft) – though production aircraft were slightly slower – while retaining the exemplary handling that had distinguished the La-5 series in combat. Despite the loss of the prototype in February 1944 to an inflight fire, the aircraft was ordered into production as the La-7. Work proceeded rapidly and the first began to reach the front as early as the late spring of 1944.

Aerial firepower

Initially sporting the twin 20mm (0.78in) ShVAK armament of the La-5FN, a new

Lavochkin La-7

Weight (Maximum take-off) 3400kg (7496lbs)
Dimensions Length: 8.6m (28ft 3in), Wingspan: 9.8m (32ft 2in), Height: 2.54m (8ft 4in)
Powerplant 1380kW (1850hp) Shvetsov M-82FN 14-cylinder air-cooled radial
Speed 661km/h (411mph)
Range 635km (395 miles)
Ceiling 10,450m (34,280ft)
Crew 1
Armament Two cowl-mounted 20mm (0.78in) ShVAK cannons or three 20mm Berezin B-20 cannons; up to 200kg (440lbs) bombload

version with three 20mm Berezin B-20 cannons mounted asymmetrically in the fuselage nose, two to port and one to starboard, was tested during June and July 1944. Offering a significantly heavier weight of fire, the three gun version went into production in January 1945, 74 being produced during that month. Shortages of the cannon led to three-gun La-7s remaining comparatively rare in service despite the armament of two ShVAK cannon being regarded as inadequate against late war German aircraft.

One of the first units to receive the new fighter was 176 Guards IAP where the La-7 would be flown by its best known pilot, Ivan Kozhedub, the most successful Allied air ace. Between his arrival with the unit on 19 August 1944 and they end of the year, Kozhedub would score 12 kills in the La-7 and on 15 February 1945 he shot down an Me 262 jet, the first destroyed over the Eastern Front.

Compared to the La-5FN, the La-7 offered usefully improved performance and handling, with its dog-fighting capability at low altitude being regarded as unbeatable. In addition, the La-7 lessened, although did not completely cure, the Lavochkin tendency to bounce on landing.

Air superiority

Although able to carry a 100kg (220lb) bomb under each wing, the La-7s exemplary air-to-air abilities meant that virtually all of the 5753 built saw service exclusively in the air superiority role. A further 582 two-seaters were built for training as the La-7UTI. As well as Soviet service, the La-7 saw brief service with Czech forces shortly before the end of the war. Czechoslovakia ultimately became the final La-7 operator, retiring theirs in 1950 – three years after its withdrawal from VVS service.

A squadron of Lavochkin La-7 pilots scramble to their aircraft. The La-7 developed into an excellent air superiority fighter.

Bereznyak-Isayev BI-1

A radical rocket-propelled point defence interceptor, the remarkable Bereznyak-Isayev BI-1 was by a considerable margin the fastest aircraft developed by the USSR during the war.

Bereznyak-Isayev BI-1

Despite never entering service, the experimental BI rocket fighters were all finished in official camouflage schemes but were devoid of individual markings.

The only rocket-propelled aircraft to enter service remains the Messerschmitt Me 163 Komet but the BI, which flew earlier and actually entered production in 1943, came extremely close to being the second. The Bereznyak-Isayev BI was the result of a proposal by Aleksander Bereznyak to mount a rocket engine in a small airframe to create a target-defence interceptor of great speed and climb rate. The power unit was the new D-1A-1100 rocket engine, designed by Aleksei Isayev and intended as an auxiliary power unit for heavy combat aircraft. The innovative nature of the project appealed to no less a personage than Josef Stalin and the rocket fighter scheme received official blessing from the very highest level.

All-wood construction

Surprisingly for such a seemingly advanced aircraft, construction was of wood throughout and thanks to its relative simplicity the first airframe was completed a mere 35 days after the first timber was cut. The D-1A-1100 engine was fuelled by kerosene and nitric acid from tanks in the nose and immediately aft of the cockpit respectively. Like the Me 163, the BI had a limited endurance of eight minutes at full power, but unlike the German aircraft the BI's motor could be throttled back to cruise, which endowed it with a relatively impressive 15 minutes powered flight time. Armament consisted of two 20mm (0.78in) ShVAK cannon with 45 rounds per gun.

Initial tests in the hands of test pilot Grigori Bakhchivandzhe progressed well, with speeds gradually increased with each successive flight. In climbing trials the BI attained an altitude of 10,000m (32,808ft) in a mere 59 seconds, an astounding performance for 1943. Unfortunately on his seventh powered flight, *Bakchivandzhe* was killed when the BI entered an uncontrolled dive at an estimated 750–800km/h (466–497mph) and broke up. Wind tunnel tests revealed that the aircraft would pitch down at high speed but revealed no obvious way to prevent this from occurring. Seven production BIs had been built by this time but work on the first batch of 50 was halted. Flight tests continued for a further two years, largely to test the rocket motor, and the programme was finally cancelled in the spring of 1945.

Bereznyak-Isayev BI-1

Weight (Maximum take-off) 1650kg (3638lbs)
Dimensions Length: 6.40m (21ft), Wingspan: 6.48m (21ft 3in), Height: 2.06m (6ft 9in)
Powerplant One 10.79kN (2,430lbf) thrust Dushkin D-1A-1100 liquid-fuel rocket motor
Speed 800km/h (497mph) (estimated)
Endurance 15 minutes
Ceiling 12,000m (39,000ft)
Crew 1
Armament Two nose-mounted 20mm (0.78in) ShVAK cannon

Curtiss P-40 Tomahawk

The P-40 was the first Allied fighter supplied to the USSR under the Lend-Lease agreement. As with other nations the Curtiss aircraft delivered reliable if somewhat unspectacular service.

Curtiss P-40 Tomahawk IIB

This ex-RAF Tomahawk IIB was flown by 154 IAP of the Red Banner Baltic Fleet during the desperate fighting around Leningrad in early 1942.

The first 20 Russian P-40s, delivered in September 1941, actually predated the Lend-Lease act and were paid for in gold. At the same time the first of an eventual 195 Tomahawk IIBs, built to British specifications, was sent as military aid by the UK. The Tomahawks were rushed into service, beginning combat operations on 12 October 1941 in the defence of Moscow with the 126th Air Defence Fighter Regiment. In a month of fighting the regiment scored 17 kills in 665 sorties.

Speed and climb

Service pilots generally rated the P-40s better than the Hurricane but inferior to the Yak-1, P-39 and German fighters, the main problems being insufficient speed and climb. On the other hand the Curtiss fighter was simple to fly, horizontal manoeuvrability was good and the aircraft was immensely strong: during one mission Aleksei Khlobystov rammed and destroyed two German aircraft, a unique achievement. The Allison engine was generally reliable but suffered in Russia from the use of low-grade oil, the same problem affecting the P-39, and the American engines were notorious among Soviet personnel for breaking connecting rods. Although many airframes were re-engined with the Klimov M-105, performance was not as good with this engine and these aircraft were relegated to secondary roles.

As well as seeing widespread use with the VVS, the Curtiss fighters also operated in the anti-shipping role with the Navy, achieving considerable success. For example, in a single day in October 1944, by which time P-40s had generally been withdrawn from front line fighter units, the 76th Regiment of the Northern Fleet destroyed two barges and six motorboats using skip-bombing tactics.

Deliveries of the P-40 continued until mid-1944 and comprised 247 P-40Bs and Cs, as well as 2178 P-40E, K, L and N models, collectively referred to in Russian service by their British names of Tomahawk and Kittyhawk respectively.

Curtiss P-40 Tomahawk IIB

Weight (Maximum take-off) 3655kg (8058lbs)

Dimensions Length: 9.66m (31ft 9in), Wingspan: 11.37m (37ft 4in), Height: 3.22m (10ft 7in)

Powerplant One 813kW (1,090hp) Allison V-1710-33 V-12 liquid-cooled piston engine

Speed 555km/h (345mph)

Range 1287km (800 miles)

Ceiling 8900m (29,100ft)

Crew 1

Armament Two 12.7mm (0.5in) Browning M2 machine guns fixed forward-firing in the nose, four 7.62mm (0.3in) Browning machine guns fixed forward-firing in the wings

Hawker Hurricane

The second most numerous fighter supplied to the Soviets, the Hurricane was not a particularly popular aircraft in Russian service but was available in quantity at a time of great need.

First flown in 1935, the Hurricane was the first modern monoplane fighter in RAF service, but by the time it was committed to combat over the Eastern Front it was well past its prime. The first Hurricanes had arrived in September 1941 with experienced British personnel charged with training Soviet pilots, although the 'instructors' did find time to fly numerous operational patrols and returned to England having shot down 16 German aircraft. These Hurricanes were handed over to naval pilots of the Northern Fleet commanded by the ace Boris Safonov to protect the vital port of Murmansk and its shipping. Hurricanes were used in force against the Finns on the Karelian Front and five Hurricane regiments took part in the Battle for Moscow.

Design flaws

Although easy to fly and extremely rugged, the Hurricane lacked the extreme responsiveness of the I-16 that most Soviet pilots were familiar with. It was also relatively slow. Hurricanes charged with escorting Pe-2s struggled to keep up with the bombers, a problem exacerbated by the lower octane rating of Russian fuel. Over half of the 2834 Hurricanes supplied to Russia were of the Mk IIb variant armed with 12 Browning 7.7mm (.303in) machine guns, which was regarded as inadequate and the majority of them were modified in the field to improve firepower. The most common replacement weapons were two 12.7mm (0.5in) UB machine guns and two 20mm (0.78in) ShVAK cannon, a process facilitated by the Hurricane's thick and spacious wing.

The final major use of the Hurricane occurred over Stalingrad and by the end of 1942 it was being withdrawn in favour of more modern types, although several front-line units soldiered on until mid-1944. As well as being pressed into service as a photo reconnaissance aircraft and glider tug, a few Hurricanes were modified with a second cockpit and a rear gun for use in the artillery-spotting role.

Hawker Hurricane IIB

Bearing 11 victory stars, Hurricane IIB 'White 80' was flown by ace Vasili Adonkin of 78 IAP in early 1943. This aircraft has been rearmed with two 20mm (0.78in) ShVAK cannon and two 12.7mm (0.5in) UB machine guns.

Hawker Hurricane IIB
Weight (Maximum take-off) 3652kg (8044lbs)
Dimensions Length: 9.83m (32ft 3in), Wingspan: 12.19m (40ft), Height: 4m (13ft 1.5in)
Powerplant 884kW (1185hp) Rolls-Royce Merlin XX V-12 liquid-cooled piston engine
Speed 528km/h (328mph)
Range 748km (465 miles)
Ceiling 10,850m (35,600ft)
Crew 1
Armament Twelve 7.7mm (0.303in) Browning machine guns; up to 450kg (1000lbs) bombload

Bell P-39 Airacobra

The Airacobra was an unusual design that proved disappointing in US and British service. By chance, it turned out to be particularly well-suited to conditions on the Eastern Front and was by far the most successful Lend-Lease fighter to serve the USSR.

Bell P-39Q Airacobra

Flown by 25-victory ace Fedor Shikunov when he was commander of 69 GvIAP, this P-39Q displays the striped rudder markings of the 23 GvIAD. Individual regiments of the division were identifiable by spinner colour.

Bell P-39Q Airacobra
Weight (Maximum take-off) 3810kg (8400lbs)
Dimensions Length: 9.19m (30ft 2in), Wingspan: 10.36m (34ft), Height: 3.78m (12ft 5in)
Powerplan One 890kW (1200hp) Allison V-1710-85 V-12 liquid-cooled piston engine
Speed 626km/h (389mph)
Range 845km (525 miles)
Ceiling 11,000m (35,000ft)
Crew 1
Armament One 37mm (1.45in) M4 cannon, four 12.7mm (0.5in) Browning M2 machine guns; up to 230kg (500lbs) bombload

The unconventional Airacobra was only the second aircraft to be built by the Bell Aircraft Corporation. Designed around its main gun, a powerful 37mm (1.45in) Colt M4 cannon, the P-39 reversed normal practice by placing the engine behind the pilot and driving the propeller via a 3m (9ft 11in) long shaft that ran between the pilot's legs. This freed up the entire nose of the aircraft for weapons, the cannon firing through the centre of the airscrew, supplemented by two Browning 12.7mm (.50in) and two 7.62mm (.30in) machine guns in the cowling. This layout was a first for a production fighter, as was the tricycle undercarriage and automobile style door employed to enter the aircraft.

Exceptional prototype performance

Powered by a turbo-supercharged Allison V-1710 V-12 engine, the prototype displayed exceptional speed and climb, reaching 6100m (20,000ft) in five minutes. However, the US Army Air Corps saw no need for a high-altitude interceptor and ordered the removal of the turbo-supercharger, crippling performance at altitude. This decision effectively sealed the aircraft's fate in US service and allowed it to be available in quantity for the Soviets.

First combat use of the Bell fighter was with the RAF but it proved a

A P-39L-1-BE Airacobra being prepared to be flown to the Soviet Union under the Lend-Lease programme, possibly at Nome, Alaska, winter 1943–44.

A pilot tends his Bell P-39 Airacobra. Most P-39s were delivered via the Alaska–Siberia ferry route.

dismal failure in British service and was rejected after a single mission, with 212 aircraft being passed on to the Soviet Union. The ex-British Airacobra Mk Is featured a 20mm (0.78in) Hispano cannon in place of the 37mm (1.45in) weapon on the centreline, retained the two 12.7mm (0.5in) machine guns in the nose but added four 7.7mm (.303in) machine guns in the wings. Arriving in December 1941, the first of these aircraft went into action in May 1942 and proved immediately effective – 153 IAP for example claimed 64 aircraft destroyed, of which 45 were fighters, in two months of operations for the loss of just eight Airacobras.

Soviet production problems

These early Airacobra deliveries were particularly timely as they coincided with a shortfall in domestic fighter production as Soviet factories were evacuated eastwards. By June 1942, new P-39Ds were beginning to be delivered directly from the US via Iran. These were fitted with the 37mm cannon, with four 7.62mm wing guns and featured provision to carry a single bomb of up to 227kg (500lb) or a drop tank under the fuselage. Ultimately almost half of all P-39 production would arrive in the USSR.

Later models differed only in detail, the final model being the P-39Q that was produced almost exclusively for Soviet needs and featured two 12.7mm (0.5in) Brownings in pods under the wings, although these were usually discarded in service to improve roll rate.

Assigned to units covering ground forces from enemy air attack, the poor altitude performance of the Airacobra was irrelevant. It was well armed, manoeuvrable, handled beautifully and was fairly fast at low altitude. Soviet pilots rated it equal or superior to contemporary models of the Messerschmitt Bf 109 and the Focke-Wulf Fw 190. But the Airacobra was also sturdy, its tricycle landing gear was particularly well-suited to rough fields at the front and in contrast to other Soviet fighters the radio was reliable. Perhaps most importantly for the averagely trained VVS pilot the Airacobra was easy to fly, and it was the combination of pleasant in-flight characteristics and combat effectiveness that led to its diminutive nickname of *Kobrushka* ('Little Cobra').

In the hands of a good pilot, the Airacobra was deadly. Grigory Rechkalov shot down 48 of his 54 confirmed 'kills' in the P-39, the most scored by any pilot flying a US-built aircraft. An Airacobra achieved the final Soviet air-to-air kill of World War II when Vasily Pshenichnikov destroyed a Fw 189 over Prague on 9 May 1945.

Bell P-63 Kingcobra

Bell's replacement for the P-39 Airacobra, the P-63 Kingcobra saw no front line service with its country of origin and over two-thirds of production was supplied to the Soviet Union.

First flown on 7 December 1942, a year to the day after the attack on Pearl Harbor, the P-63 retained the unusual fuselage layout of the P-39 and mated it to a laminar flow wing. Despite an overall similarity to the earlier aircraft, including retention of such idiosyncratic features as the 'car-door' cockpit access, the P-63 was an entirely new aircraft, somewhat larger in all dimensions. It featured the same armament of two 12.7mm (.50in) Browning machine guns and one 37mm (1.4in) Oldsmobile cannon firing through the propeller spinner, augmented by two further 12.7mm guns in underwing fairings.

Unlike the P-39, it featured a second supercharger for high-altitude work and performance was generally excellent at all altitudes. However, with sufficient numbers of P-51s and P-47s in service it was clear that the US would not require the P-63 and, cognisant of the success of the P-39 over the Eastern Front, further development was sanctioned largely with Russian use in mind.

Bell P-63C Kingcobra

Probably on the strength of 16 GvIAP in 1945, this P-63 was finished in all-over olive drab. The ventral fin identifies this as a P-63C.

Pacific operations

Production deliveries began in October 1943 and 3303 Kingcobras would ultimately roll out of Bell's factory. Of these, 2421 were flown via Canada and Alaska across the Bering Sea to the Soviet Union. Officially, Soviet use of the Kingcobra as agreed with the US was restricted to the Pacific area of operations, the type scoring its first confirmed victory on 15 August 1945 during the brief Soviet campaign against Japan, when Lieutenant I. F. Miroshnichenko shot down a Ki-43 Hayabusa.

However, there is some evidence that P-63s were secretly used against Germany in the closing days of the war in Europe. As the most potent Lend-Lease fighter supplied to the USSR, the P-63 was retained in frontline VVS service until 1951, receiving the NATO reporting name of FRED.

Bell P-63C Kingcobra

Weight (Maximum take-off) 4853kg (10,700lbs)

Dimensions Length: 9.96m (32ft 8in), Wingspan: 11.68m (38ft 4in), Height: 3.84m (12ft 7in)

Powerplant One 1300kW (1800hp) Allison V-1710-117 V-12 liquid-cooled piston engine

Speed 660km/h (410mph)

Range 720km (450 miles)

Ceiling 13,000m (43,000ft)

Crew 1

Armament One 37mm (1.45in) M10 cannon, four 12.7mm (0.5in) Browning M2 machine guns; up to 680kg (1500lbs) bombload

Supermarine Spitfire

Britain's premier fighter was not ideally suited to conditions on the Eastern Front and was destined to play a relatively short-lived role on operations.

The first Supermarine Spitfires to be received by the USSR were a group of 150 Mk VBs that started to arrive during the Spring of 1943. These aircraft had been personally requested by Stalin in October of the previous year and were drawn from RAF reserve stocks in the Middle East, most having seen extensive service. Soviet pilots were not overly impressed with the aircraft and although its handling and easy flying characteristics were appreciated, its performance was found to be inferior to contemporary Yaks and Lavochkins. It was also known that the markedly superior Mk IX was in production and service, and there was some resentment that the VVS had only been provided with these 'old' second-hand Spitfires.

Despite their misgivings, in combat the British fighter gained respect for its armament, which was the heaviest of any Soviet single seater, and for its all-metal construction that was found to be more resilient to fire than the largely wooden Yaks. The Spitfire V saw brief but intensive service with Soviet units before being replaced by Airacobras, a decision that would have amazed Western pilots.

Supermarine Spitfire Mk IX UTI

No two-seat Spitfires were supplied to the Soviets but many were subsequently modified into trainers. This Spitfire IX UTI was converted by 1 Aircraft Depot Leningrad in 1945.

Limited combat role

In February 1944 the first of an eventual 1185 Spitfire Mk IXs began to arrive in Russia, which in contrast to the VBs were highly regarded. Assigned to protect major cities and strategic targets due to their excellent climb and altitude performance, they saw virtually no action as by this stage of the war very few German aircraft were penetrating deep into Soviet airspace. The only known combat occurred on 8 March 1945 when two Spitfire IXs shot down a Junkers Ju 88 near Leningrad at a height beyond that attainable by any other Soviet fighter. Retained in service in the intercept role, its unrivalled altitude performance resulted in the Mk IX remaining operational in Soviet service as late as 1951.

Supermarine Spitfire Mk IX

Weight (Maximum take-off) 4309kg (9500lbs)

Dimensions Length: 9.47m (31ft 1in), Wingspan: 11.23m (36ft 10in), Height: 3.49m (11ft 5in)

Powerplant One 1175kW (1575hp) Rolls-Royce Merlin 66 V-12 liquid-cooled piston engine

Speed 657km/h (408mph)

Range 698km (434 miles)

Ceiling 13,106m (43,000ft)

Crew 1

Armament Two 20mm (0.8in) Hispano Mk.II cannons and four 12.7mm (0.5in) Browning Mk.II machine guns fixed forward firing in wings

Republic P-47 Thunderbolt

Despite the P-47's excellent reputation with other Allied air forces, the Soviet Union was not impressed with this massive fighter and found little use for the Thunderbolt.

Republic P-47D Thunderbolt

This P-47D was serving on anti-shipping duties with 255 IAP of the Northern Fleet in late 1944. Soviet forces found little use for the massive Republic fighter.

Soviet authorities had expressed an interest in Republic's P-47 as early as 1943 and ultimately 195 Lend-Lease P-47Ds would arrive in the USSR. Oddly, one further P-47D was supplied to the Soviet Union, which was uniquely paid for privately by funds raised by American senators and was the only 'razorback' example of the fighter to appear in Russia.

Aerodynamic shortcomings

Although the installed armament of eight 12.7mm (.50in) Browning machine guns was viewed favourably, test flights of the Thunderbolt, optimized for high altitude combat, were disappointing. In marked similarity to the earlier MiG-3, it was found to be wanting in manoeuvrability and possessed poor acceleration at the medium and low altitudes where virtually all combat occurred over the Eastern Front. At the same time the aircraft was studied in great detail at the Bureau of New Technology of the People's Commissariat of the Aviation Industry, particularly the turbo-supercharger installation. Soviet technicians found much to praise in terms of the aircraft's build quality, which was considerably superior to Soviet aircraft at the time although they were far less impressed with its aerodynamics.

As a result of its unsuitability as a 'frontal' fighter, the Thunderbolt was primarily assigned to air defence duties such as protecting strategic targets, and as a result saw no action as German long-range aviation activity was effectively over by this time.

The Soviet Navy also expressed an interest in the P-47 as an anti-shipping aircraft and carried out tests of dive bombing and bombing from mast height with a variety of bombloads. The tests were deemed successful and the Thunderbolt went into service with the 255th IAP of the Northern Fleet in the winter of 1944, operating in this role for about a year.

Republic P-47D
Weight (Maximum take-off) 7938kg (17,500lbs)
Dimensions Length: 11.02m (36ft 1.2in), Wingspan: 12.43m (40ft 9in), Height: 4.47m (14ft 8in)
Powerplant One 1500kW (2000hp) Pratt & Whitney R-2800-59 18-cylinder air-cooled radial piston engine
Speed 686km/h (426mph)
Range 1660km (1030 miles)
Ceiling 13,000m (42,000ft)
Crew 1
Armament Eight 12.7mm (0.5in) M2 Browning machine guns; up to 1100kg (2500lbs) of bombs

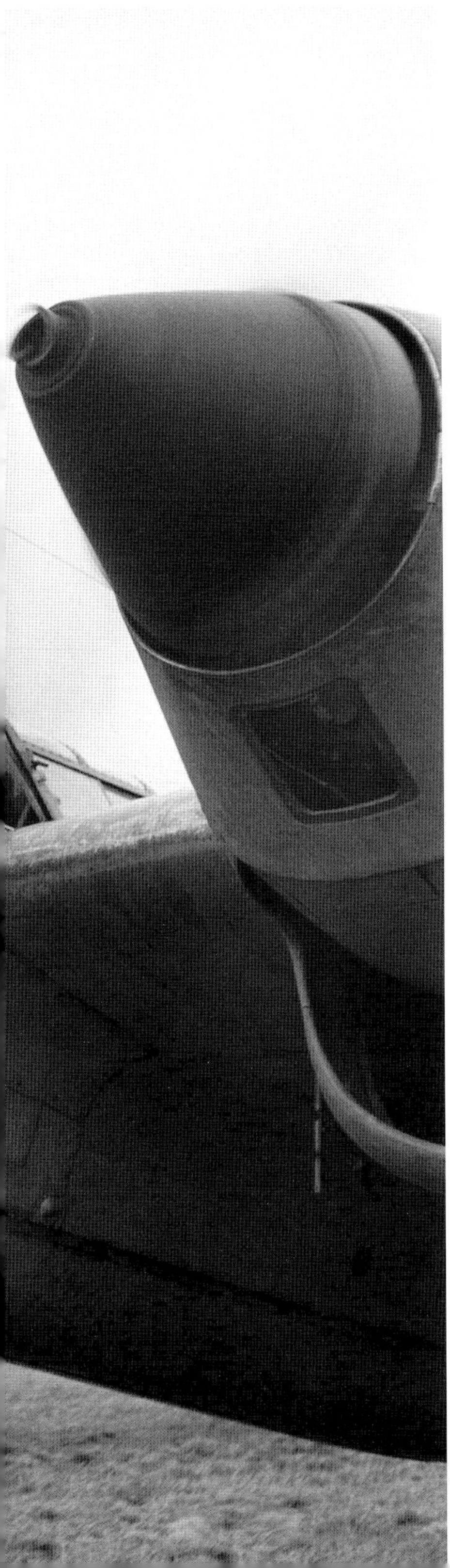

BOMBERS AND GROUND-ATTACK AIRCRAFT

The majority of Soviet bombers were tactical aircraft, and over the battlefield the VVS became an immensely powerful ground-attack force. It included the uniquely Soviet night harassment bombers, whose effect was intended to be psychological rather than material. The relatively few strategic bombers were organized into the *Aviatsiya Dal'nevo Deystiva* ('Long-Range Aviation'). The following aircraft are included in this chapter:

- Tupolev TB-3
- Polikarpov Po-2
- Tupolev SB-2
- Yakovlev UT-1
- Ilyushin DB-3
- Sukhoi Su-2
- Ilyushin Il-4
- Arkhangelsky Ar-2
- Yakovlev Yak-2
- Yakovlev Yak-4
- Petlyakov Pe-2
- Yermolaev Yer-2
- Ilyushin Il-2
- Petlyakov Pe-8
- Tupolev Tu-2
- Douglas A-20
- Handley-Page Hampden
- North American B-25

A captured Tupolev SB in Finnish service, October 1943. The Finns salvaged many SBs after they had crash-landed in Finland during the Winter War of 1939–40.

Tupolev TB-3

The world's first four-engined cantilever monoplane heavy bomber, the TB-3 was effectively obsolete at the outbreak of hostilities, yet performed valuable work during the initial phases of the conflict.

Essentially a stretched TB-1, the TB-3 was first flown in December 1930 and entered service during 1932. The TB-3 was an impressive performer, setting several payload-to-altitude records and achieving a record endurance of over 18 hours. About half the production aircraft were fitted with Mikulin M-17 engines but latter production TB-3s utilized the more powerful Mikulin AM-34. Given the stately performance of the TB-3, early M-17 aircraft struggling to exceed 200km/h (124mph), it is not surprising that the defensive armament provided was comprehensive (by mid-thirties standards at least), featuring twin gun positions in the nose and two amidships as well as two underwing 'dustbin' turrets with single gun each, all with drum-fed Degtyaryov DA 7.62mm (0.3in) machine guns.

Underwing turrets

Later aircraft featured a tail position in place of the underwing turrets and switched to the ShKAS 7.62-mm gun that were often turret mounted. Bombload throughout its life remained a maximum of 5000kg (11,000lbs).

Despite having been officially retired in 1939, the VVS still had over 500 operational TB-3s at the time of the German attack. Being based well to the east of the front line, the TB-3s escaped the widespread destruction and were pressed back into frontline service. Daylight raids proved near-suicidal in the lumbering TB-3, so a switch was made to nocturnal operations, although not before several had been employed as the mothership for I-16 parasite dive bombers to successfully attack targets in Romania. Throughout the first two years of war the TB-3s were heavily committed, crews sometimes flying three missions a night and the aircraft was also used in combat as a paratroop transport.

Transport duties

The transport role became this rugged aircraft's main function after 1943. Of the 818 built, 10 were still operational in July 1945.

Finnish troops examine a Tupolev TB-3 bomber which has been captured following an emergency landing in March 1940.

Tupolev TB-3

Weight (Maximum take-off) 19,300kg (42,549lbs)

Dimensions Length: 24.4m (80ft 1in), Wingspan: 41.8m (137ft 2in), Height: 8.5m (27ft 11in)

Powerplant Four 526kW (705hp) Mikulin M-17F V-12 liquid-cooled piston engines

Speed 212km/h (132mph)

Range 2000km (1200 miles)

Ceiling 4800m (15,700ft)

Crew 8–10

Armament Two paired 7.62mm (0.3in) Degtyaryov DA machine guns flexibly mounted in nose, two paired 7.62mm (0.3in) DA machine guns flexibly mounted in port and starboard dorsal gun positions, one 7.62mm (0.3in) DA machine gun in ventral 'dustbin' turret; up to 5000kg (11,000lbs) bombload

Tupolev TB-3

Operational during the summer of 1942, this TB-3 of 250 APDD has been modified with two MV-3 turrets on the staggered dorsal positions, each armed with a single 7.62mm (0.3in) ShKAS machine gun.

Polikarpov Po-2

The most produced biplane in history, the Po-2 served the USSR in huge numbers as a trainer yet also operated as a night harassment bomber to immense psychological effect.

A typical aircraft of its era, Nikolai Polikarpov designed the Po-2 to replace the U-1, a copy of the Avro 504, as the standard Soviet basic trainer. Flown for the first time on 24 June 1927, the prototype was overweight, slow and handled badly. A major redesign cured the problems and the U-2, as it was then known, entered service as a basic trainer. Its excellent short takeoff and landing (STOL) performance led to its widespread use as a liaison aircraft and the U-2S (for *Sanitarniy*, 'Ambulance') carried a stretcher for medical evacuation.

In 1935 production of the U-2VVS was begun, which was an armed version for the Army Co-operation role, 468 of which were built by 1936. In addition to its military service, the U-2 was used as an agricultural aircraft from the start, leading to its nickname of *Kukuruznik*, a slang term for a farm worker, from the Russian *kukuruza,* 'maize'. To the Germans though, the Polikarpov was the *Nähmaschine* 'sewing machine' due to the distinctive engine sound, and its female pilots were the feared *Nachthexen*, 'Night Witches'.

Light night bomber

The U-2's enduring wartime fame would derive from its use as a bomber, initially field conversions of training aircraft, later the purpose-built U-2LNB, LNB standing for *Lyoghkiy Nochnoiy Bombardirvshchik* or 'Light Night Bomber', carrying up to 300kg (661lb) in bombs on underwing racks and armed with a ShKAS machine gun in the rear cockpit. Flown from their bases to jumping-off airfields at dusk, the U-2s missions were very short range

Polikarpov Po-2 U2

Weight (Maximum take-off) 1350kg (2976lbs)

Dimensions Length: 8.17m (26ft 10in), Wingspan: 11.4m (37ft 5in), Height: 3.1m (10ft 2in)

Powerplant One 93kW (125hp) Shvetsov M-11D 5-cylinder air-cooled radial piston engine

Speed 150km/h (93mph)

Range 630km (390 miles)

Ceiling 3000m (9800ft)

Crew 2

Armament One 7.62mm (0.3in) ShKAS machine gun flexibly mounted in rear cockpit; up to six 50kg (110lbs) bombs

Polikarpov Po-2 U2

Wearing typical mid-war three-colour camouflage, this U-2 'White 9' was on the strength of 46 ZAP in July 1943.

and many sorties could be made by the same aircraft, an extreme example being that of U-2 crew Yekaterina Ryabova and Nadezhda Popova of the famed all-female 588th Night Bomber Regiment, who on one occasion flew 18 missions in a single night.

Although these raids caused little material damage, their effect on enemy troops cannot be overstated. Approaching near silently, with the engine throttled back or even, on occasion, switched off, the Polikarpov biplanes could strike without warning at any time and the maddening effect of the sleep deprivation this caused was profound. U-2 attacks were practically unstoppable: a small target, flying at night, at treetop height or lower, the U-2's 150km/h (93mph) maximum speed was slower than the stalling speed of modern fighters rendering it virtually impossible to intercept.

Its wood and canvas construction was almost invisible to radar and flak was largely ineffective, German gunners finding that firing on a U-2 served only to give their position away and often led to attack. So successful were the Polikarpovs that the *Luftwaffe* started pressing its own obsolete biplane fighters and trainers into *Nachtschlachtgruppen* 'Night attack groups' in direct imitation of the Soviet nocturnal harassment units.

Two members of the famous all-female 588th Night Bomber Regiment stand in front of their Polikarpov Po-2 bomber. Known as the 'Night Witches', the unit was renowned for low-altitude night attacks on Axis ground positions.

Renamed

After Nikolai Polikarpov's death in 1944 the U-2 was renamed Po-2 in his honour and remained in production in the USSR until 1952. Post-war, the Po-2 would again see combat as a nocturnal nuisance bomber, gaining fame as 'Bedcheck Charlie' during the Korean War and becoming the only biplane known to have destroyed a jet fighter, albeit accidentally: an F-94 Starfire crashed attempting to intercept a low-flying Po-2 at night.

Meanwhile civil use of the Po-2 by flying clubs, as a glider tug and crop-sprayer, continued for decades. The last Po-2s are believed to have been built in Poland in 1959 but it is possible later examples were assembled from spare parts.

The total number of U-2/Po-2s constructed is unknown but is believed to be between 30,000 and 40,000.

Tupolev SB-2

Despite being one of the world's most advanced bombers when it entered service in 1936, such was the frenetic pace of development in the late thirties that the SB-2 was obsolete by the time of the German invasion.

The SB-2's origin lay with a requirement for a bomber aircraft fast enough to be able to operate without a fighter escort, a proposal that was seized upon enthusiastically by Andrei Tupolev and his associate Aleksander Arkhangelsky. They drew up a sleek twin engine bomber consisting of the smallest possible airframe able to carry the specified bombload of 500kg (1102lb).

The aircraft, known initially as the ANT-40, was to feature a duralumin-stressed skin construction, the first Soviet aircraft so constructed, and was as aerodynamically clean as possible.

Performance

Fitted with two imported 544kW (730hp) Wright Cyclone radials the prototype achieved a top speed of 325km/h (202mph) at 4000m (13,123ft), only 5km/h (3.1mph) less than that stipulated in the specification. An engine change in the second prototype to two 581kW (779hp) Hispano-Suiza 12Ybrs V-12 engines led to a jump in performance to a spectacular 430km/h (267mph), exceeding the specification by an undeniably impressive 100km/h (62mph).

Speed reduction

The addition of military equipment and an increase in wing and tailplane area reduced the speed slightly on the first production aircraft, now designated SB-2 (the 'SB' standing for *Skorostnoy Bombardirovschik* 'High speed Bomber'), although performance remained extremely good overall. The first aircraft were fitted with the Klimov M-100, a licence-built version of the Hispano-Suizas fitted to the prototype and was capable of 393km/h (244mph).

Up to 500kg (1102lb) of bombs could be carried and the aircraft was defended by four 7.62mm (0.3in) ShKAS machine guns, two in the

Tupolev SB-2

Weight (Maximum take-off) 7880kg (17,372lbs)

Dimensions Length: 12.57m (41ft 3in), Wingspan: 20.33m (66ft 8in), Height: 3.6m (11ft 10in)

Powerplant Two 716kW (960hp) Klimov M-103 V-12 liquid-cooled piston engines

Speed 450km/h (280mph)

Range 2300km (1400 miles)

Ceiling 9300m (30,500ft)

Crew 3

Armament Two 7.62mm (0.3in) ShKAS machine guns flexibly mounted in nose, one 7.62mm ShKAS machine gun flexibly mounted in dorsal position and one 7.62mm ShKAS machine gun flexibly mounted in ventral hatch; six 100kg (220lbs) or six 50kg (110lbs) bombs in bomb-bay, two 250kg (550lbs) bombs on wing racks

Tupolev SB-2

This SB-2 of an unknown unit force-landed behind German lines in August 1941. It is in the standard pre-war temperate scheme of dark green over light blue.

nose and one each in dorsal and ventral positions.

Production began in the spring of 1936 and service introduction was rapid – by September of the same year the SB-2 entered combat over Spain where it was initially immune from interception due to its great speed, although losses rose once the Messerschmitt 109 was introduced to the conflict, a portent of things to come. Of the 93 SBs supplied to the Spanish Government, 19 survived the conflict serving until 1948.

Licensed version

During 1937 the design was licensed by Czechoslovakia and 101 were produced there as the Avia B-71; these aircraft were absorbed into the *Luftwaffe* after the annexation of Czechoslovakia in 1938, with some subsequently being passed on to Bulgaria.

In July of the same year the Sino-Japanese War broke out and SB-2s were operated by China, with further conflict with Japan occurring during the Khalkhin Gol battles of 1939. By the time of the Winter War with Finland the SB-2's speed advantage had been effectively erased and it was vulnerable to fighter attack – at least 100 were lost during 15 weeks of conflict.

Over the course of 1940 the design had been aerodynamically improved, most noticeably featuring new engine nacelles with the bluff frontal radiators replaced with a cleaner underslung arrangement. This, coupled with a move to the M-103 engine from late 1938, increased speed and the defensive armament, while the remaining four ShKAS machine guns were improved by providing an MV-3 turret for the dorsal weapon and an MV-2 ventral mount both of which offered much improved fields of fire. The bombload by now had increased to a maximum of 1500kg (3307lb).

VVS bomber

SB-2 production actually increased in 1940 due to the absence of any viable replacement and it was by far the most numerous VVS bomber when Germany invaded, with between 1500 and 2000 deployed along the Western border of the USSR. Many were destroyed on the ground and terrible losses were incurred desperately attempting to stem the German advance before the aircraft largely gave way to the Pe-2.

A switch to nocturnal operations occurred after only a few days and the aircraft was heavily employed attacking German positions by night during the defence of Moscow and Leningrad in the latter half of 1941.

After 1942 the SB was mostly employed in second line duties as a trainer and glider tug, although surprisingly the fighting career of the SB-2 was not yet over.

Chinese Civil War

Immediately after VJ day in 1945, Nationalist Chinese forces flew SB-2s in action against Communist forces during the Chinese Civil War, the aircraft being finally retired in 1946.

A three-man crew of a Tupolev SB-2 prepare for a mission. The SB-2 was the mainstay of the Soviet bomber force in 1941, but sustained such heavy losses that it was phased out of frontline duties in 1942.

Yakovlev UT-1

Alexander Yakovlev's UT-1 was designed as an advanced trainer and sport aircraft, although an armed conversion was briefly manufactured as a light attack aircraft.

Closely related to the UT-2, the single seat UT-1 was initially designated AIR-14, 'AIR' in honour of Alexei Ivanovich Rykov, although when Rykov fell from favour and was executed in 1938 the designation was hastily swapped to UT-1. First flown in early 1936, the aircraft was intended as an advanced fighter trainer. At the time of its introduction air combat training was accomplished on retired fighter aircraft such as the Polikarpov I-5, which was not ideal as these aircraft were both obsolete and expensive to operate. Powered similar to the UT-2 with the economical M-11 radial, the UT-1 offered a cheap, modern solution.

In production from 1937, the UT-1 was noted for its excellent aerobatic qualities but also for the very precise handling it demanded from its pilot. A spate of pre-war crashes led to the suspicion that the aircraft was dangerous and its use was strictly limited, and in 1940 the UT-1 was declared 'not suitable for training cadets'. Production ceased after 1241 had been built and it was generally expected to be withdrawn during 1941.

Nocturnal attack

The outbreak of war changed the fortunes of the UT-1 and it was widely employed as a communications and liaison machine with frontline units. Acute shortages of more suitable aircraft also led to its employment in combat roles.

Reconnaissance missions were flown by the UT-1 from the summer of 1941 but the unarmed aircraft was vulnerable. It was decided that it could be used as a light nocturnal attack aircraft and a series of around 100 factory conversions were built.

Yakovlev UT-1B
Weight (Maximum take-off) 759kg (1673lbs)
Dimensions Length: 5.78m (18ft 10in), Wingspan: 7.30m (23ft 11in), Height: 2.34m (7ft 7in)
Powerplant One 82kW (110hp) Shvetsov M-11D 5-cylinder air-cooled radial piston engine
Speed 190km/h (118mph)
Range 520km (323 miles)
Ceiling 4600m (15,092ft)
Crew 1
Armament Two 7.62mm (0.3in) ShKAS machine guns fixed forward-firing on wing; four RS-82 rockets on underwing racks

Yakovlev UT-1

A typical early-war UT-1. Spats were generally discarded in service, especially on those aircraft operating from primitive frontline fields where mud would often clog the wheels within their fairings.

Designated the UT-1B, it featured two over-wing ShKAS machine-guns and four RS-82 rockets on racks. Mainly used in the Caucasus region, the UT-1Bs saw action over Stalingrad and into early 1943 but were soon replaced when Il-2s became available. Many UT-1s survived the war and were prized by civilian flying clubs for their aerobatic capabilities.

YAKOVLEV UT-2

Yakovlev's monoplane basic trainer was built in large numbers, of which a considerable amount saw use as light night harassment bombers.

Derived from the pre-war AIR-9 light aircraft, the UT-2 dispensed with more complicated elements such as cockpit canopy, flaps and slats to produce an exceptionally utilitarian trainer. Originally flown as the AIR-10 during May 1935, the aircraft was accepted for production in September 1937. Its simple construction aided mass production and by 1941 it had usurped the Polikarpov U-2 as the standard Soviet basic trainer, remaining the most widespread equipment of VVS flying schools until 1948. A few early aircraft were fitted with the Renault Bengali engine but the vast majority used the readily available M-11, as fitted to the U-2. Even with the modest power of this engine the performance was good, and pre-war UT-2s set international records in speed, range and altitude for its weight class. Despite its training role, the UT-2 was not a particularly easy aircraft to fly and was prone to entering flat spin. This was remedied by sweeping back the wings and modifying the control services slightly on the UT-2M of 1941.

In the desperate fighting following the launch of Operation Barbarossa, it was decided that the UT-2 should be used as a light night bomber. Modifications were made at field level and it is believed that around 2000 were used in this role during the war. The redesign involved the removal of the pupil's cockpit and often two ShKAS machine guns were fitted on top of each wing, firing outside the propeller arc. Up to 300kg (6614lb) of ordnance could be carried as underwing stores, such as PTAB anti-tank bomblets, conventional bombs or up to four RS-82 rockets. Heavily employed from the summer of 1941 onwards, by late 1943 the UT-2s were sidelined by more powerful aircraft and switched to providing air support to partisan groups.

Ilyushin DB-3

After demonstrating spectacular long-range capability in propaganda flights in the late 1930s, on 7 August 1941 the DB-3 became the first Soviet aircraft to bomb Berlin.

Developed from the all-wood TsKB-26, the first Soviet twin engine aircraft to be looped, the DB-3 was an all-metal long-range bomber that, although it could carry five times the bombload of Tupolev's SB-2, was only 24km/h (15mph) slower. It also boasted such modern features as self-sealing fuel tanks and a crew intercom system, both included as a result of Soviet aircrew experience over Spain.

The crew of three consisted of the pilot, navigator/front gunner with a 7.62mm (0.3in) ShKAS and the radio operator/rear gunner who was required to divide his time between the dorsal turret and a ventral gun position, both armed with a single ShKAS. The maximum bombload was 2500kg (5511lb), which was the heaviest of any twin engine bomber at the time the DB-3 entered service in 1937.

New construction techniques

The aircraft featured new construction techniques that allowed for a light structure but was difficult to build and production was slow – only 45 aircraft being built by two factories during 1937, delaying the planned replacement of the Tupolev TB-3 in long-range units. Eventually 1528 were

built before production switched to the Il-4 in 1939. After seeing combat over China and Finland, the DB-3 was heavily engaged in fighting the German invaders from June 1941. Initially used by day, losses of the lightly defended DB-3 were considerable. In a single raid on 26 June 1941 for example, 43 DB-3s were lost. By 1942 the aircraft had wisely switched to nocturnal operations.

In naval service the DB-3T variant was employed as a torpedo bomber in the anti-shipping role and remained in frontline service for the duration of the war, attacking Japanese shipping in Manchuria as late as August 1945.

Ilyushin DB-3

An M-86 engined DB-3T of 2 MTAP serving with the Black Sea Fleet in the spring of 1941. Black Sea Fleet DB-3Ts were the first Soviet aircraft to bomb Berlin in August 1941.

Ilyushin DB-3TP (Prototype)

The DB-3TP floatplane was converted from a standard DB-3 with the addition of a pair of British-built floats. The aircraft could carry a 45-36-AN or 45-36-AV torpedo or a 900kg (1984lb) mine. Tested at Sevastopol during 1938, the aircraft was found unsatisfactory and no production ensued.

Ilyushin DB-3

Weight (Maximum take-off) 9450kg (20,834lbs)
Dimensions Length: 14.22m (46ft 8in), Wingspan: 21.44m (70ft 4in), Height: 4.19m (13ft 9in)
Powerplant Two 709kW (951hp) Nazarov 9-cylinder air cooled radial engines
Speed 439km/h (273mph)
Range 3800km (2400 miles)
Ceiling 9600m (31,500ft)
Crew 3
Armament One 7.62mm (0.3in) ShKAS machine gun flexibly mounted in nose, one 7.62mm (0.3in) ShKAS machine gun in dorsal turret, one 7.62mm (0.3in) ShKAS flexibly mounted in ventral hatch; up to 2500kg (5511lbs) bombload

Sukhoi Su-2

The first aircraft designed by the great Pavel Sukhoi, the Su-2 was a light bomber that proved vulnerable in its intended role but enjoyed greater success as a reconnaissance platform.

Sukhoi Su-2

This Su-2 wears the pre-war scheme of overall green over blue. 'Red 2' was on the strength of 226 BAP on the Southwestern Front in July 1941.

Originally designated ANT-51 as Sukhoi was then part of the Tupolev design bureau, the first prototype was slower than the specification demanded but the design as a whole was considered sound and a switch from the 610kW (818hp) Shvetsov M-62 engine to the 746kW (1000hp) Tumansky M-87 improved matters, delivering a top speed of 468km/h (291mph) at 5600m (18,373ft). In this form the aircraft was accepted for production as the BB-1, BB standing for *Blizhniy Bombardirovshchik* 'Short range Bomber'.

Armament

Up to 400kg (882lb) of bombs could be carried internally, although the aircraft also sported racks for up to ten RS-82 or RS-182 rockets and four 7.62mm (0.3in) ShKAS machine-guns in the wings. Defensive armament consisted of two further ShKAS weapons, one mounted in the MV-5 rear turret and a further weapon firing through a ventral hatch. However, the M-87 engine was proving unreliable and a further change to the Tumansky M-88 did not massively improve matters as this engine was also suffering teething issues. The final few Su-2s of the 910 constructed were built with the reliable Shvetsov M-82 and were blessed with impressive performance, but the decision had already been made to terminate production by this time.

Reconnaissance role

In operational service the Su-2's offensive armament was considered too light and its defensive weapons inadequate, the cannon-armed, armoured Il-2 and the larger, faster Pe-2 being preferred for the attack and light bomber roles. Despite its relatively minimal armour and reputation for vulnerability, the Su-2 was sturdily constructed, able to absorb a great deal of punishment and still survive. After the Battle of Stalingrad in late 1942 the Su-2 was mainly employed for reconnaissance, a role in which it excelled, serving in this capacity until early 1944.

Sukhoi Su-2
Weight (Maximum take-off) 4620kg (10,185lbs)
Dimensions Length: 14.30m (46ft 11in), Wingspan: 14.30m (46ft 11in), Height: 3.75m (12ft 4in)
Powerplant One 992kW (1330hp) Shvetsov M-82 14-cylinder two-row air-cooled radial engine
Speed 468km/h (291mph)
Range 910km (565 miles)
Ceiling 8400m (27,560ft)
Crew 2
Armament Four 7.62mm (0.3in) ShKAS fixed in wings, one 7.62mm (0.3in) ShKAS in rear turret; up to 400kg (882lbs) bombload or 10 RS-82 rockets

Ilyushin Il-4

Starting life as the DB-3F, the Il-4 represented a complete redesign and after initial engine problems it served effectively for the duration of the Great Patriotic War.

Ilyushin Il-4

This Il-4 from an unknown unit, possibly 815 BAP, appeared in a Soviet newsreel of 1944. The basic black-and-green camouflage has been field modified with light brown.

The DB-3 had given the VVS its first modern all-metal long-range bomber but it was difficult to build and had proved somewhat unreliable in service. Ilyushin sought to eradicate these problems while improving performance with the DB-3F. First flown by famed test pilot Vladimir Kokkinaki on 21 May 1939, the aircraft was effectively a totally new design featuring a completely rebuilt structure, greatly influenced by the latest American construction methods introduced to the Soviet Union with licence production of the Douglas DC-3. The time required to build the DB-3F was 14,331 man hours compared with 30,301 hours for the original DB-3; ultimately this would drop to 12,500 hours, which was even fewer than that required to construct the smaller Pe-2. Performance was improved by the adoption of a thinner wing of greater area and the fitment of two Tumansky M-88 engines in place of the M-87s of the DB-3B.

Initially the M-88 suffered from overheating issues but these were eventually overcome, although not before the first series of DB-3Fs was completed with the older engine. In an attempt to alleviate the inadequate defensive armament, a fourth crew member was added allowing a dedicated gunner for each of the three weapons and an MV-3 turret in the dorsal position that allowed greater freedom of movement for the gunner. All three weapons remained the same 7.62mm (0.3in) ShKAS machine guns as previous variants, although from the spring of 1942 Il-4s – as the aircraft was re-designated on 26 March of that year – replaced the ShKAS with a 12.7mm (0.5in) Berezin UBT-armed UTK-1 turret.

Despite bombing Warsaw, Königsberg and Danzig the day after hostilities began, the DB-3Fs were thrown into a desperate defence against the German invaders (much like virtually every aircraft facing the onslaught of Operation Barbarossa). Operating in a tactical role, for which they had not been designed, and required to fly by day – which rendered them vulnerable to fighters – early losses were severe. However, there were many available and the airfields where these long-range bombers were based were sited well behind the front line and not one was destroyed on the ground in the opening attack.

By late 1941 missions were invariably flown at night, Berlin becoming a regular destination for the bombers from August of that year and Il-4s maintained a largely successful bombing campaign until the end of the

Ilyushin Il-4

Weight (Maximum take-off) 12,120kg (26,720lbs)

Dimensions Length: 14.76m (48ft 5in), Wingspan: 21.44m (70ft 4in), Height: 4.10m (13ft 6in)

Powerplant Two Tumanksy M-88B 14-cylinder air cooled radial engines

Speed 410km/h (250mph)

Range 3800km (2400 miles)

Ceiling 8700m (28,500ft)

Crew 4

Armament One 7.62mm (0.3in) ShKAS machine gun flexibly mounted in nose, one 12.7mm (0.5in) Berezin UBT machine gun in dorsal turret, one 7.62mm (0.3in) ShKAS machine gun flexibly mounted in ventral hatch; up to 2700kg (6000lbs) of bombs or mines or one 940kg (2100lbs) Type 45-36 torpedo

war. In production until mid 1945, the Il-4 was quickly withdrawn from service thereafter, disappearing from bomber units during 1946. Like the DB-3T, the Il-4 was also developed into a torpedo bomber, the Il-4T, and operated with considerable success both in the West and in brief but intense action in the Pacific during August 1945. The Il-4T outlived its conventional bomber counterpart by a considerable margin, the last retiring in 1952.

Arkhangelsky Ar-2

The final iteration of Tupolev's redoubtable SB to enter production, the Arkhangelsky Ar-2 was an aerodynamically improved version that was optimized for dive bombing.

Arkhangelsky Ar-2

This aircraft, probably of 33 BAP, 19 BAD, was found abandoned by advancing German troops at Belaya Tserkov, Gorodishche, Kiev district, in the summer of 1941.

Arkhangelsky Ar-2
Weight (Maximum take-off) 8150kg (17,967lbs)
Dimensions Length: 12.50m (41ft), Wingspan: 18m (59ft), Height: 3.56m (11ft 8in)
Powerplant Two 820kW (1100hp) Klimov M-105R V-12 liquid cooled piston engines
Speed 505km/h (313mph)
Range 1000km (621 miles)
Ceiling 10,500m (34,400ft)
Crew 3
Armament One 7.62mm (0.3in) ShKAS machine gun flexibly mounted in nose, one 7.62mm (0.3in) ShKAS machine gun flexibly mounted in dorsal position, one 7.62mm (0.3in) ShKAS flexibly mounted in ventral hatch; up to 1500kg (3306lbs) bombload

After Andrei Tupolev's arrest in October 1937, Alexander Arkhangelsky had taken over the leadership of his design bureau and oversaw the development of a new, high-speed dive bomber SB variant, initially designated the SB-RK. Featuring redesigned outer wing panels with increased taper, better streamlined engine nacelles with the radiators now buried in the wing rather than in a chin position and a new and more aerodynamic nose, the SB-RK was expected to attain 600km/h (373mph) at 6000m (19,685ft).

To fulfil the dive bombing requirement, a PBP-1 bombsight was fitted for the pilot and a new bomb rack was developed. Recovery from the dive was automatically controlled after the bombs were released. Built during 1940, testing began during October and although the speed fell short of expectations, only managing a disappointing maximum of 512km/h (318mph), the handling was found to be improved over the SB-2 and the dive-bombing equipment functioned flawlessly. Production was authorized and the first examples of the aircraft, now known as the Ar-2, were built at Zavod 22 from December 1940.

Production of the Ar-2 ceased in March 1941 only three months after it had started, and Zavod 22 switched to building the Pe-2. Of the 198 Ar-2s delivered, many were destroyed on the ground in the opening few days of Operation Barbarossa. The survivors were worked hard and losses were heavy, 95 being recorded as combat losses during 1941. Like most Soviet bombers the Ar-2's defensive armament was light and although fairly fast it was vulnerable to enemy fighters. Compared to the Pe-2, Arkhangelsky's dive bomber was easier to fly, particularly on take-off and was considered more forgiving to novice pilots but performance was not in the same league. Most were destroyed or withdrawn during 1942 although a handful survived in frontline service into 1943.

Yakovlev Yak-2

The Yak-2 caused a sensation when it first flew in 1939 and much was expected of the aircraft in service. Those expectations were never realized and the Yak-2 faded into obscurity.

Designed by Aleksander Yakovlev as the Ya-22, the aircraft was intended to be the basis for a cannon-armed twin-engine fighter, short-range bomber and a tactical reconnaissance aircraft all utilizing the same basic airframe. From the start of the project, the overriding intention was to maximize performance by reducing weight and drag to the lowest possible level and in this regard the aircraft was very successful.

During testing in 1939 the first prototype recorded a maximum speed of 567km/h (352mph) at 4900m (16,076ft), making it one of the fastest military aircraft in the world at the time, a fact reinforced by a dramatic appearance at that year's May Day parade over Red Square.

Rushed design

Impressed by the prototype's speed, Stalin ordered production to begin as the BB-22 *Blizhniy Bombardirovshchik* 'Short-range Bomber' before the prototype had completed testing. This was gravely premature as problems with engines overheating soon arose, the undercarriage had to be strengthened and the flaps enlarged. The aircraft was both laterally and longitudinally unstable and it was unpopular, not least because escape in the event of an emergency was practically impossible. Worst of all was the inevitable weight growth caused by the installation of military equipment and modifications. This meant that when coupled with poor-quality workmanship on production BB-22s, the aircraft could only manage 478km/h (297mph) at 4600m (15,092ft) – much the same as the earlier SB-2, which could carry twice the bombload nearly three times greater distance. In service, crews joked that BB stood for *Bespoleznyy Bombardirovshchik* 'Useless Bomber'.

Redesignated Yak-2 in December 1940, production was terminated at the 111th example and combat use, primarily as a reconnaissance platform, was very brief but extremely intensive. Most were destroyed during the first month of Operation Barbarossa – of 73 Yak-2s on strength on 22 June 1941, only one remained airworthy by 18 July.

Yakovlev Yak-2

Commander of 316 RAP, Lt.I.M. Agarkov, flew Yak-4 'Yellow 33' on the Southwestern Front during July–August 1941.

Yakovlev Yak-2

Weight (Maximum take-off) 5660kg (12,477lbs)
Dimensions Length: 9.34m (30ft 8in), Wingspan: 14m (45ft 11in), Height: 4.18m (13ft 9in)
Powerplant Two 716kW (960hp) Klimov M-103 V-12 liquid-cooled piston engines
Speed 478km/h (297mph)
Range 800km (500 miles)
Ceiling 8900m (29,200ft)
Crew 3
Armament One 7.62mm (0.3in) ShKAS machine gun fixed forward-firing in nose, one ShKAS machine gun flexibly mounted in rear cockpit; up to 600kg (1300lbs) bombload

Yakovlev Yak-4

A developed Yak-2, the Yak-4 offered little better performance than its progenitor and only 90 were produced.

Keen to improve upon the underwhelming Yak-2, Yakovlev reengined the aircraft with a pair of more powerful 821kW (1101hp) Klimov M-105s in place of the 715kW (959hp) M-103s. The troublesome cooling system was redesigned, easily distinguishable externally by prominent oil coolers in the chin position under each engine. Fuel tanks were added in the wings, conferring a useful increase in range. Armament remained the same: two 7.62mm (0.3in) ShKAS machine guns as the Yak-2 but the rear gun position was altered to improve the field of fire, although crew accommodation was still considered to be cramped.

The bombload was increased to 900kg (1980lbs), though even this relatively modest load had a negative effect on the aircraft's centre of gravity, leading to difficult handling qualities – a problem that had also afflicted the earlier aircraft.

Tests compromised

Replacing the Yak-2 on the production line, the first Yak-4s (then known as the BB-22bis), began to appear in October 1940. However, a mere 90 examples would be constructed by the time production ceased on 11 February 1941 as the aircraft in tested form was still not considered combat capable. Given the success of the considerably more formidable Pe-2 no case could be made to continue development of the Yak twin.

Nonetheless, the 90 Yak-4s were delivered to frontline units, the final 21 arriving in April 1941 leaving little time for crew familiarization before the opening of hostilities on 22 June. In combat the Yak-4 was found to be fast and crews appreciated the fact it did not burn easily, although they were less pleased with the small bombload and light armament. Like the Yak-2, the Yak-4 was heavily committed to action and suffered accordingly, virtually all being destroyed by the end of 1941 although a single example survived with the 118th Reconnaissance regiment unit the end of the war.

Yakovlev Yak-4

The few Yak-4s to see service were generally finished in anonymous schemes. 'Yellow 3' was destroyed on the ground at Grodno airfield, Belarus, in June 1941.

Yakovlev Yak-4

Weight (Maximum take-off) 6115kg (13,481lbs)
Dimensions Length: 10.18m (33ft 5in), Wingspan: 14m (45ft 11in), Height: 4.18m (13ft 9in)
Powerplant Two 821kW (1101hp) Klimov M-105 V-12 liquid-cooled piston engines
Speed 533km/h (331mph)
Range 960km (595 miles)
Ceiling 9700m (31,800ft)
Crew 2
Armament One 7.62mm (0.3in) ShKAS machine gun fixed forward-firing in nose, one ShKAS machine gun flexibly mounted in rear cockpit; up to 900kg (1980lbs) bombload

Petlyakov Pe-2

Tough, manoeuvrable and very fast, the Pe-2 was one of the finest tactical aircraft of World War II.

The Pe-2 was originally conceived as a turbo-supercharged high altitude fighter with a pressurized cockpit designated the VI-100, which first flew in December 1939 and demonstrated excellent performance. However, Soviet Authorities believed that high altitude bombers did not constitute a serious threat, which was correct as it turned out. Cognizant of the success of German *Blitzkreig* tactics in the West and impressed by the dive-bombing capability of the Junkers Ju 88, Soviet authorities demanded Petlyakov convert the VI-100 into a dive bomber, allowing a mere 45 days to effect this major conversion.

New dive bomber

Amazingly the deadline was met, the new aircraft dispensing with turbo-superchargers and pressurisation. A fuselage bomb bay and two small bomb bays in the rear of the engine nacelles, glazed nose, dive brakes, defensive armament and a third crew member were also added.

The new aircraft was perceived to be so important that tooling up for mass production was initiated before the first example had even flown. Designated the PB-100 (PB standing for *Pikiruyushchiy Bombardirovshchuk*, 'dive Bomber') the first production aircraft flew on 15 December 1940. The designation was changed to Pe-2 in the same month and the first Pe-2-equipped units were formed during March 1941.

In its initial form, the Pe-2 was an impressive performer. Its level speed was identical to the Bf 109E, its excellent manoeuvrability belied its fighter origins and it was capable of carrying an internal bombload of 600kg (1323lb), plus a further 400kg (882lb) externally. In a dive-bombing attack the nose glazing allowed the pilot to visually acquire the target before initiating the dive, recovery being automatic. The aircraft mounted two 7.62mm (0.3in) ShKAS machine guns in the nose, later production models replacing one of these weapons with a 12.7mm (0.5in) Berezin UBK. Defensive armament consisted of two trainable

Petlyakov Pe-2

With its winter camouflage nearly worn off, this early-production Pe-2 'White 9' of 46 BAP was operational in the Moscow area during the dark days of the winter of 1941–42.

Petlyakov Pe-2

Weight (Maximum take-off) 8495kg (18,728lbs)

Dimensions Length: 12.66m (41ft 6in), Wingspan: 17.16m (56ft 4in), Height: 3.5m (11ft 6in)

Powerplant Two 903kW (1211hp) Klimov M-105PF V-12 liquid-cooled piston engines

Speed 580km/h (360mph)

Range 1160km (720 miles)

Ceiling 8800m (28,900ft)

Crew 3

Armament One 7.62mm (0.3in) ShKAS machine gun and one 12.7mm (0.5in) Berezin UB fixed forward-firing in the nose, one 12.7mm (0.5in) Berezin UB machine gun in rear of turret, one Berezin UB flexibly mounted in ventral hatch, one 7.62mm (0.3in) ShKAS interchangeable between port and starboard window mounts; up to 1000kg (2205lbs) bombload

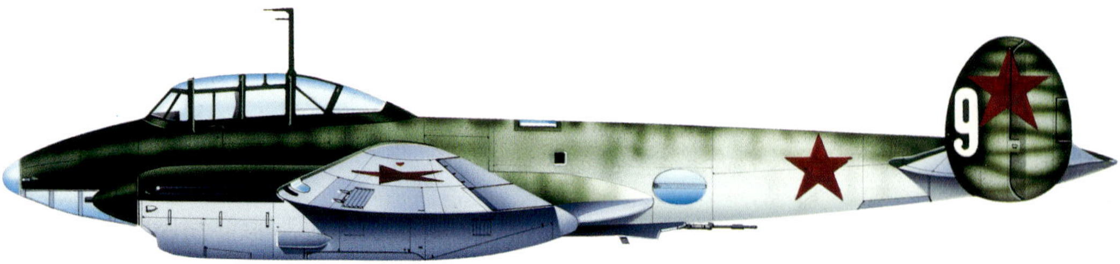

The clean lines of the Pe-2 are clearly evident in this view of Soviet Air Force Pe-2FTs. Fighter escorts were seldom provided for Pe-2 bombers after 1942.

7.62mm ShKASs in dorsal and ventral positions. These weapons were later replaced by 12.7mm UBKs, the dorsal weapon mounted in a VUB-1 turret on later aircraft. An additional ShKAS was provided that could fire out of small beam windows in the rear fuselage, the gunner moving the weapon to one side or other as required.

Performance issues

Service entry of the Pe-2 proved taxing. The aircraft was complicated, build quality was poor and performance suffered. Take-off and landing characteristics were difficult, the pilot requiring considerable physical strength in order to lift off – a shortcoming that was particularly problematic for female pilots. Pilots then had to deal with a high landing speed, a propensity to stall and an incipient bounce on touch down. Still a new aircraft when the German invasion began, inexperienced Pe-2 crews suffered high attrition. Successes were achieved even at this early stage though, a raid on the Romanian Ploesti oilfield set alight a quarter of a million tons of oil and Pe-2 formations were successful in retaliating Bf 109 attacks. As production continued incremental changes were made on the production line, the most significant being the switch to M-105PF engines during 1942, and build quality improved. By mid 1943 the Pe-2 was a thoroughly proven aircraft and the most numerous multi-engine Soviet bomber.

A reconnaissance variant known as the Pe-2R was also developed, this variant gaining some 500km (310 miles) of range with the use of two 290 litre (64 gallons) drop tanks. The capricious handling issues led to a trainer version, the Pe-2UT, being built from September 1943. Concerns about the availability of the M-105 engine led to the construction of a Pe-2 powered by the Shvetsov M-82 radial. This version entered production in August 1943. Ultimately only 32 examples were built as the M-82 was urgently required for the La-5 but 24 of these aircraft did see service, operating mostly with reconnaissance units during late 1943 and 1944.

Active service

Production eventually ceased in December 1945 although the aircraft served with several air forces after VJ day, the final examples being retired by Poland in 1957 by which time the aircraft had received the NATO reporting name 'Buck.' Over 11,000 Pe-2s were built, making it the third most produced twin engine aircraft of the war after the Junkers Ju 88 and Vickers Wellington.

Petlyakov Pe-2FT

This Pe-2FT was flown by Lieutenant Colonel Vasili I. Rakov, commander of the 12th Guards Bomber Air Regiment in the Gulf of Finland in 1944, who was twice awarded the Gold Star of a Hero of the Soviet Union.

Rear Gunner
A gunner manned a 12.7mm (0.5in) Berezin UB machine gun in the rear of the cockpit for defensive purposes.

Nose Armament
The Pe-2 mounted two 7.62mm (0.3in) ShKAS machine guns in the nose, later production models replacing one of these weapons with a 12.7mm (0.5in) Berezin UBK.

Bombload
The Pe-2 was capable of carrying an internal bombload of 600kg (1323lb) and a further 400kg (882lb) externally.

Petlyakov Pe-2FT

Weight (Maximum take-off) 8495kg (18,728lbs)
Dimensions Length: 12.66m (41ft 6in), Wingspan: 17.16m (56ft 4in), Height: 3.5m (11ft 6in)
Powerplant Two 903kW (1211hp) Klimov M-105PF V-12 liquid-cooled piston engines
Speed 580km/h (360mph)
Range 1160km (720 miles)
Ceiling 8800m (28,900ft)
Crew 3
Armament One 7.62mm (0.3in) ShKAS machine gun and one 12.7mm (0.5in) Berezin UB fixed forward-firing in the nose, one 12.7mm (0.5in) Berezin UB machine gun in rear of turret, one Berezin UB flexibly mounted in ventral hatch, one 7.62mm (0.3in) ShKAS in dorsal position; up to 1000kg (2205lbs) bombload

Yermolaev Yer-2

Despite being built in relatively small numbers, the Yermolaev Yer-2 was an important design that saw intensive combat.

Yermolaev Yer-2

An M-105 engined Yer-2 of 421 BAP in late 1942 or early 1943, with a black painted underside for its nocturnal role. This aircraft was flown by Lt. Colonel A.G. Gusev, commander of the unit.

The Yermolaev Yer-2 was derived from a highly advanced prototype airliner – the Stal-7 – designed by Italian-born engineer Roberto Ludvigovich Bartini, development of which had passed to his deputy, Vladimir Yermolaev, when the luckless Bartini had been imprisoned on espionage charges in 1938. Under Yermolaev the airliner was redesigned as the DB-240, 'DB' standing for *Dalny Bombardirovshcik* 'Long-range Bomber', first flying on 14 May 1940.

Klimov engines

Performance was not quite as good as expected due to the aircraft being forced to use two Klimov M-105s rather than the higher-powered M-106s it was designed for, as the M-106 was suffering from insuperable cooling problems. On the other hand, the DB-240 offered an excellent range with a 1000kg (2205lb) bombload, the defensive armament was considered good and the aircraft entered production as the Yer-2 in March 1941.

Wartime priorities

The Yer-2's initial production life was short, with only 128 constructed when in August 1941 Zavod 18 switched to building the more urgently required Il-2. In the same month the Yer-2 entered combat for the first time when on the 10th three examples bombed Berlin, one subsequently being shot down in error by I-16s on its return to Soviet airspace.

Later committed to interdicting the German advance on Moscow, the Yer-2s suffered high losses in low-level tactical missions they were ill-suited for but remained in frontline service until August 1943 when the last survivors were transferred to training units.

New version

Development of the aircraft continued and Yer-2 production was reinitiated towards the end of 1943, now equipped with efficient Charomskiy ACh-30 diesel engines. Production difficulties delayed service entry and the first bombing mission of the 'new' Yer-2 occurred on 7 April 1945, the aircraft remaining in service until the late 1940s.

Yermolaev Yer-2

Weight (Maximum take-off) 18,580kg (40,962lbs)

Dimensions Length: 16.42m (53ft 10in), Wingspan: 23m (75ft 6in), Height: 4.82m (15ft 10in)

Powerplant Two 1100kW (1500hp) Charomskiy ACh-30B V-12 liquid-cooled diesel piston engines

Speed 507km/h (315mph)

Range 5500km (3417 miles)

Ceiling 7200m (23,600ft)

Crew 4

Armament One 12.7mm (0.5in) Berezin UBT flexibly mounted in nose, one 12.7mm Berezin UBT on flexible ventral mount, one 20mm (0.78in) ShVAK cannon in dorsal turret; up to 5000kg (11,023lbs) bombload (short-range bombing only)

Ilyushin Il-2 *Shturmovik*

Described by Stalin 'as essential to the Red Army as air and bread', the heavily armed and armoured Il-2 *Shturmovik* remains the most produced combat aircraft of all time and the most iconic Soviet aircraft of World War II.

Ilyushin Il-2

During 1945 the Il-2s of 56 GvShAP participated in making a film called *Victory*. For the film, the inscription 'Stalingrad-Berlin' was painted over the existing fuselage markings.

Combat over Spain had revealed the vulnerability of conventional aircraft attacking ground forces and a specification was issued for a dedicated, armoured, ground-attack aircraft. In response, Sergey Ilyushin designed the TsKB-55, a two seat, single engine design in which the entire forward structure consisted of an armoured steel *bronekorpus* ('armour body') surrounding the engine and crew.

Single seat conversion

First flown on 2 October 1939, the TsKB-55 proved to be overweight and underpowered. As a weight-saving measure the aircraft was converted into a single seater; to improve handling the wings were swept back by five degrees and the tailplane area was increased. To address visibility concerns the engine was moved forward and lowered by 175mm (6.9in) while the pilot's seat was raised. The AM-35 engine was switched for the AM-38 that delivered maximum power at lower altitude, and cannon armament was fitted, initially two 20mm (0.78in) ShVAKs although many production aircraft would be fitted with the 23mm (0.9in) Volkov-Yartsev VYa-23. An abrupt partially glazed fairing covered the deleted second cockpit lending the aircraft a distinctive profile and earning it the nickname *gorbatiy* – 'hunchback'. In this form the Il-2 entered production, the first machine off the assembly line flying on 10 March 1941.

Inexperienced crew

The Il-2 had barely entered service when Operation Barbarossa was launched in June 1941. Up to 249 Il-2s were nominally available but the aircraft was so new that flight manuals had not been written nor had any significant training been undertaken by crews. Typical was the experience of the 4th ShAP, the first unit to take the Il-2 into combat, where not a single pilot had even fired the Il-2's guns before they were thrust into action on 27 June. It quickly became apparent that the Il-2 was particularly vulnerable to fighter attack from the rear, 60 per cent of

Ilyushin Il-2

Weight (Maximum take-off) 6360kg (14,021lbs)

Dimensions Length: 11.65m (38ft 3in), Wingspan: 14.60m (47ft 11in), Height: 4.17m (13ft 8in)

Powerplant One 1280kW (1,720hp) Mikulin AM-38F liquid-cooled V-12 engine

Speed 410km/h (250mph)

Range 765km (475 miles)

Ceiling 4525m (14,846ft)

Crew 2

Armament Two 23mm (0.9in) VYa-23 cannons and two 7.62mm (0.3in) ShKAS machine guns fixed forward firing in wings, one 12.7mm (0.5in) Berezin UBT machine gun flexibly mounted in rear cockpit; up to 600kg (1320lbs) bombs in wing bomb-bays and underwing racks, or four dispensers for 48 2.5kg (5.5lbs) PTAB anti-armour bombs (192 total) in wing bomb-bays; eight RS-82 rockets or four RS-132 rockets

Ilyushin Il-2

Il-2s featured more unofficial artwork and slogans than most VVS aircraft, usually on the fuselage. This eagle-adorned two-seater was operating with an unknown unit in Latvia during 1944.

all losses being attributed to fighters during 1941–42, and urgent steps were taken to address this shortcoming. At least 1000 Il-2s were field modified with an extemporized gun position, usually featuring one or two 7.62mm (0.3in) ShKAS machine guns, this being accomplished by simply cutting a hole in the fuselage immediately behind the cockpit canopy and attaching a gun mount.

A factory built two-seat Il-2 entered production in September 1942 featuring a Berezin UBT 12.7mm (0.5in) gun for a gunner who sat on a simple strap in the rear fuselage with some relatively thin armour and a sideways opening canopy, although this was often discarded in service to improve the field of fire. By summer 1943 losses to fighters had fallen to 37 per cent but at considerable cost: rear gunner casualties were seven times higher than those of pilots. Two-seat Il-2s were in action from 30 October 1942 and by the end of the year around 1450 two seaters had been built, some examples being equipped with 37mm (1.45in)

From mid-1941, Il-2s were equipped for the use of the RS-132 rocket on the production lines, although RS-82 weapons are being carried by aircraft '8' in this winter 1941/42 shot.

***Shturmovik* Tactics: Wheel of Death**

The 'Circle of Death' could be highly effective against panzer units caught in the open. The circle allowed each pilot to pick a target, before attacking in turn from a shallow dive with rockets, bombs or cannon fire. It also provided a fair degree of protection against marauding *Luftwaffe* fighters as each aircraft covered the one in front. Repeated attacks would be made until the *Shturmoviks* had fired off all their ammunition.

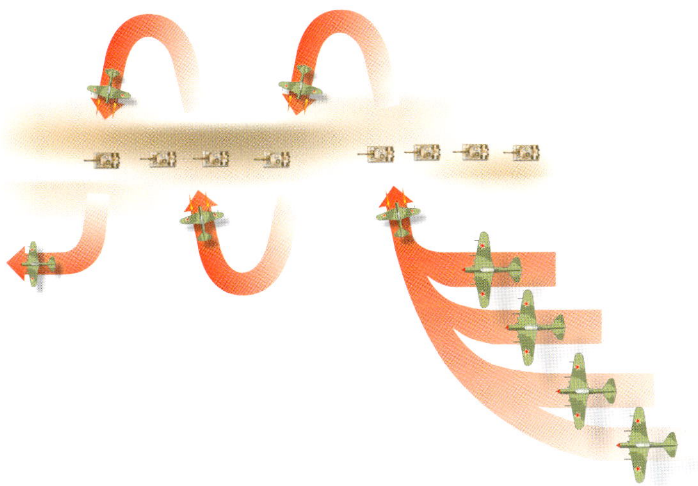

***Shturmovik* Tactics: Attacking a column**

Il-2s in 'snake' combat formation attacking an armoured column. The aircraft approach in peleng (staggered line abreast) formation at an altitude of 800–1200 metres (2620–3940ft) moving into line astern before diving at 30°–40° to fire rockets, bomb or strafe. By 'snaking' across the length of the column, rather than making straight attack runs, the aircraft are a far more difficult target for AA fire.

Nudelman-Suranov NS-37 cannon. The effective PTAB *protivotankovaya aviabomba* – 'anti-tank aviation bomb' – shaped-charge 2.5kg (5.5lb) bomblets were added in 1943, of which the Il-2 could carry 192.

In addition to these weapons the Il-2 could carry up to 600kg (1323lb) of conventional bombs or eight RS-82 or RS-132 rockets. In September of the same year the Il-2's wing was further swept to 15 degrees to address the rearward shift of centre of gravity caused by the twin seat redesign. Known as *Strielka* – 'Arrow' – and often referred to as the Il-2M3, although no such designation existed in contemporary Soviet records, this became the definitive form of the Il-2 and was produced until October 1945.

'Wheel of Death'

In action, Il-2 tactics were honed to a high degree of effectiveness, usually flown at very low level – so much so that on occasion *Shturmoviks* flew so low they were obscured by their own dust cloud. Against armour Il-2 units employed a formation known as the 'Wheel of Death' – aircraft would circle their targets then attack in a shallow dive before rejoining the circle and repeating assaults until all their ordnance had been expended. Il-2s also served with naval units as a torpedo bomber and even achieved

Il-2 *Shturmovik*

This Il-2 was on strength with the 566th Attack Aviation Regiment, 277th Attack Aviation Division, Soviet Frontal Aviation, and served on the Leningrad front in summer 1944. Flown by squadron commander V.I. Mykhlik, the inscriptions on the fuselage read 'For Leningrad' and 'Revenge for Khristenko', a pilot lost in combat in early 1944.

Defensive Gun
With the introduction of a second crewman, a 12.7mm (0.5in) UBT machine gun was added on a flexible mount in the rear cockpit.

SHTURMOVIKS AT KURSK

During the Battle of Kursk in the summer of 1943, Vasily Ryazanov, commander of the 1st Attack Aviation Corps, made notably effective use of the Il-2 as part of a combined-arms force that used the *Shturmoviks* alongside infantry, armour and artillery. Prior to the battle, the Soviets assembled 35 per cent of their entire combat aircraft strength at Kursk, and these enjoyed superior maintenance and larger forward fuel reserves than their *Luftwaffe* counterparts. The Il-2s operated over Kursk in large numbers, harassing German armour and racking up huge victory claims, although their own losses were significant. Ryazanov twice received the Gold Star of Hero of the Soviet Union, and his 1st Attack Aviation Corps became the first unit to be awarded the honorific title of Guards. The victory at Kursk stands as a turning point in the war, after which the Red Army held the strategic initiative until the end of the war.

Ilyushin Il-2

Weight (Maximum take-off) 6360kg (14,021lbs)

Dimensions Length: 11.65m (38ft 3in), Wingspan: 14.60m (47ft 11in), Height: 4.17m (13ft 8in)

Powerplant One 1280kW (1720hp) Mikulin AM-38F liquid-cooled V-12 engine

Speed 410km/h (250mph)

Range 765km (475 miles)

Ceiling 4525m (14,846ft)

Crew 2

Armament Two 23mm (0.9in) VYa-23 cannons and two 7.62mm (0.3in) ShKAS machine guns fixed forward firing in wings, one 12.7mm (0.5in) Berezin UBT machine gun flexibly mounted in rear cockpit; up to 600kg (1320lbs) bombs in wing bomb-bays and underwing racks, or four dispensers for 48 2.5kg (5.5lbs) PTAB anti-armour bombs (192 total) in wing bomb-bays; eight RS-82 rockets or four RS-132 rockets

Cockpit Armour
The pilot was seated in an armoured tub, with a thickness of 5–12mm (0.2–0.5in) and which also extended to protect the engine, and was capable of defeating all small arms fire.

Rockets
This aircraft is seen unleashing RS-132 unguided rockets from the underwing racks. These weapons were introduced in mid-1941 and were used together with the earlier RS-82. A total of six RS-132s could be carried, later increased to eight.

Engine
The Il-2 was powered by a Mikulin AM-38F liquid-cooled V-12, developing 1280kW (1720hp). Introduced in July 1942, it was capable of running on motor fuel.

some success in the air-to-air role, its performance rendering it able to successfully engage most German bombers and transport aircraft.

The Il-2 was quite possibly the most important aircraft to serve the VVS – over 36,000 were built and at least 10,000 lost in combat. In addition to the USSR, Il-2s served with Czech, Polish and Yugoslavian forces during the war, with Mongolia, Hungary and Bulgaria flying them after 1945. Post-war the aircraft received the NATO reporting name 'Bark', the final Il-2s being retired by both Bulgaria and Yugoslavia in 1954.

Ilyushin Il-2

Flown by 3 ShAP 'Ostrava Regiment' of the Czechoslovakian Mixed Air Division in the spring of 1945, this was a presentation aircraft with the name 'Chapaevksy' painted on both sides of the fuselage, honouring the workers of Chapaevsk. The same aircraft stands at the front of the row of Il-2s in the photograph below.

Stalin threw his weight behind the Il-2 project. When production delays were occurring, he said that 'The Red Army needs the Il-2 as it needs air or bread...this is my final warning.'

ILYUSHIN IL-10

Despite resembling its illustrious Il-2 forebearer, the Il-10 was a completely new aircraft. Too late to see major service in World War II, it enjoyed a lengthy post-war career. Somewhat surprisingly, the Il-10's origins lay in a proposal for a single-seat, heavily armoured fighter intended to attack bombers and transports. When air force enthusiasm for such an aircraft waned, Ilyushin reworked the design as a two seat ground attack aircraft to replace the Il-2. During tests to decide a successor to the Il-2, the Il-10 beat two other contenders: a second Ilyushin design, the Il-8, and the rival, radial-engined Sukhoi Su-6.

Committed to production in August 1944, the Il-10 featured exactly the same armament as the Il-2 and followed the same concept of a heavily armoured airframe. Powered by a 1320kW (1770hp) Mikulin AM-42 fitted to a smaller airframe, the Il-10 was around 145km/h (90mph) faster than the older aircraft and possessed almost fighter-like manoeuvrability.

End of war action

Entering service with operational units in January 1945 and in action from 15 April 1945, Il-10s were flown by three units before the end of hostilities around three weeks later. A single unit also operated the Il-10 against rail and shipping targets in the brief Soviet campaign against Japan during 1945. After VJ day the Il-10 saw further combat during the Korean war of 1950–53 and in Chinese service against Taiwan in 1955. After 1945 the Il-10 replaced the Il-2 as the standard *Shturmovik* aircraft in Soviet service. Later models featured a heavier armament of four 23mm (0.9in) Nudelman-Suranov NS-23 cannon with a 20mm (0.787in) Berezin B-20 cannon for the rear gunner. Production lasted until 1954 in the USSR although the Il-10 was also built under licence in Czechoslovakia as the Avia B-33 until 1956. The last operational Il-10s were retired by China in 1972.

Ilyushin Il-10

Weight (Maximum take-off) 6535kg (14,407lbs)

Dimensions Length: 11.06m (36ft 3in), Wingspan: 13.40m (43ft 11in), Height: 4.18m (13ft 9in)

Powerplant One 1320kW (1770hp) Mikulin AM-42 liquid-cooled V-12 engine

Speed 551km/h (342mph)

Range 800km (500 miles)

Ceiling 5500m (18,000ft)

Crew 2

Armament Two 23mm (0.9in) Vya-23 cannon and two 7.62mm (0.3in) ShKAS machine guns fixed forward firing in wing, one 20mm (0.78in) Berezin B-20 cannon or one 12.7mm (0.5in) Berezin UBT machine gun flexibly mounted in rear turret; up to 600kg (1320lbs) bombs in wing bomb-bays and underwing racks, or four dispensers for 48 2.5kg (5.5lbs) PTAB anti-armour bombs (192 total) in wing bomb-bays; eight RS-82 rockets or four RS-132 rockets

Ilyushin Il-10

'Red 1' served with Soviet Occupation Forces, probably 2 GvShAP, in Germany in 1950. The overall aluminium finish was unusual, even in peacetime.

Ilyushin Il-10

The Il-10 had a lengthy post-war career: this Il-10M dates from 1954, by which time the aircraft was armed with four 23mm (0.9in) cannon in the wings and had garnered the NATO reporting name 'Beast'.

Petlyakov Pe-8

The only modern four-engined bomber developed by the USSR, the Pe-8 was an impressive performer sidelined by almost total Soviet fixation on tactical aircraft.

Petlyakov Pe-8

This AM-35A engined Pe-8 was based at Kratovo airfield, near Moscow, in March 1943. It served with 746 APDD. A Pe-8 flew from this base to drop the first 5000kg (11,000lb) bomb on Königsberg in May of the same year.

Intended as a replacement for the venerable TB-3, the aircraft that would become the Pe-8 flew for the first time as the ANT-42 on 27 December 1936. After some modification the aircraft was approved for production as the TB-7, the designation being changed to Pe-8 to honour its chief designer, Alexander Petlyakov.

The first four production aircraft featured a complicated system wherein a 750kW (1006hp) Klimov M-100 was housed in the fuselage and drove the supercharger supplying the four Mikulin AM-34FRN engines. Difficult to manufacture, this system was abandoned and much of the subsequent development involved trying to find a suitable engine; although only 93 were built, four different engines were fitted to production aircraft.

After several were built with the AM-35A engine, attention moved to the Charomskiy M-30 diesel that offered exceptional range capability but proved unreliable in service. The last Pe-8s featured Shvetsov M-82 radials with better power but lesser range than the diesels.

Long-range bomber

Although always a rare aircraft, there never being more than 30 operational examples in service, the Pe-8 was used throughout the war and beyond on very long-range missions, bombing Berlin for the first time on 10 August 1941. Early in its service life the Pe-8 had a very good performance and was difficult to intercept, the loss rate in 1942 being only one per 106 missions but as the war progressed, German defences improved and losses more than doubled by 1944.

Unlike most contemporary Soviet aircraft, the Pe-8 was well known in the West after an AM-35A powered version carried Foreign Minister Molotov on a much photographed flight to Washington DC via the UK in 1942, attracting considerable attention.

Petlyakov Pe-8

Weight (Maximum take-off) 35,000kg (77,162lbs)

Dimensions Length: 23.2m (76ft 1in), Wingspan: 39.13m (128ft 5in), Height: 6.2m (20ft 4in)

Powerplant Four 999kW (1340hp) Mikulin AM-35A V-12 liquid-cooled piston engines

Speed 443km/h (275mph)

Range 3700km (2300 miles)

Ceiling 9300m (30,500ft)

Crew 11

Armament Two 7.62mm (0.3in) ShKAS machine guns in nose turret, one 12.7mm (0.5in) Berezin UBT in port and starboard inner engine nacelle turret, one 20mm (0.78in) ShVAK in dorsal turret, one 20mm (0.78in) ShVAK in rear turret; up to 5000kg (11,000lbs) bombload

Tupolev Tu-2

Designed by Andrei Tupolev while he was serving a 10-year prison sentence for sabotage and espionage activity, the Tu-2 was a superb aircraft that enjoyed a long production and service life.

The disgraced Tupolev's name was not attached to the project and the new bomber was referred to simply as 'Aircraft 103', with Tupolev's prison design office being 'Department 103'. The three-seat aircraft was designed to an impossibly ambitious specification, a maximum level speed of 700–740km/h (435–460mph) and ceiling of 13,000–14,000m (42,6501–45,932ft) were both specified, and the bomber was to be capable of both level and dive bombing with a bombload of up to 2000kg (4409lb). Construction of Aircraft 103 began in May 1940 and the first flight took place on 29 January 1941.

Despite not meeting the unrealistic specification, the streamlined Tu-2 demonstrated excellent performance: powered by two Mikulin 1119kW (1500hp) AM-37 inline engines with a maximum speed of 635km/h (395mph) being recorded at 7800m (25,590ft), which was faster than virtually all operational fighters at the time. A major change to the design instigated almost immediately was inspired from a rather unusual source; impressed by an early example of a Junkers Ju 88 bought in 1940, the number of crew was increased to four, which required substantial alteration of the cockpit. Although not as finely streamlined as the prototype, the '103U' as it was now

Tupolev Tu-2S

Flown by A. Pushkin, the commander of 88 BAD, this Tu-2S was operational during the Soviet advance in the spring of 1945.

dubbed was only marginally slower. Unfortunately the AM-37 engines were proving unreliable and the only readily available alternative was the Shvetsov M-82 radial, which – though proven and dependable – led to a further reduction in maximum speed to 528km/h (328mph). Nonetheless, the M-82 powered 103V was approved for production a mere five days before the German invasion, and a month later Tupolev was released from detention. The aircraft was officially renamed Tu-2 on 28 March 1942.

Tu-2S model

Despite being assessed as the finest medium bomber available, the Tu-2 was not produced in numbers until the last year of the war. During 1942 there was a critical shortage of fighters and Tu-2 production was terminated in October after 80 had been constructed, the factory switching to building Yak-9s instead. Although not in serial production, development of the aircraft continued and reports from Tu-2 crew at the front were enthusiastic. Production restarted in the summer of

Tupolev Tu-2S

Weight (Maximum take-off) 11,768kg (25,944lbs)

Dimensions Length: 13.8m (45ft 3in), Wingspan: 18.86m (61ft 11in), Height: 4.13m (13ft 7in)

Powerplant Two 1380kW (1850hp) Shvetsov ASh-82 14-cylinder air-cooled radial piston engines

Speed 528km/h (328mph)

Range 2020km (1260 miles)

Ceiling 9000m (30,000ft)

Crew 4

Armament Two fixed forward-firing 20mm (0.78in) ShVAK cannon in the wings, three 12.7mm (0.5in) flexibly mounted rear-firing Berezin UBT machine-guns in rear cockpit canopy, dorsal and ventral hatches; up to 3000kg (6600lbs) bombload

1943 of the 1380kW (1850hp) Shvetsov ASh-82FN-powered TU-2S. In service from early 1944 and used almost exclusively in the level bombing role, the Tu-2S was popular and effective.

Easy to handle even with the maximum bombload of 3000kg (6600lb), the Tu-2 was fast enough to be difficult to intercept. It was well armed, with three 12.7mm (0.5in) Berezin UBT machine-guns in two dorsal and one ventral positions, in addition to the two fixed 20mm (0.78in) ShVAK cannon in the wing roots.

Post-war the Tu-2 was produced until 1948, by which time 2527 had been built, which saw further combat in the Chinese Civil War and over Korea.

Tupolev Tu-2S

Seen here is a Tupolev Tu-2S wearing a similar camouflage scheme to aircraft which participated in the Soviet Aviation day parade over Moscow on 18 August 1945.

Tupolev Tu-2S

Weight (Maximum take-off) 11,768kg (25,944lbs)
Dimensions Length: 13.8m (45ft 3in), Wingspan: 18.86m (61ft 11in), Height: 4.13m (13ft 7in)
Powerplant Two 1380kW (1850hp) Shvetsov ASh-82 14-cylinder air-cooled radial piston engines
Speed 528km/h (328mph)
Range 2020km (1260 miles)
Ceiling 9000m (30,000ft)
Crew 4
Armament Two fixed forward-firing 20mm (0.78in) ShVAK cannon in the wings, three 12.7mm (0.5in) flexibly mounted rear-firing Berezin UBT machine-guns in rear cockpit canopy, dorsal and ventral hatches; up to 3000kg (6600lbs) bombload

---

Defensive Armament
The Tu-2 included three hand-held 12.7mm (0.5in) Berezin UBT machine-guns in two dorsal and one ventral positions.

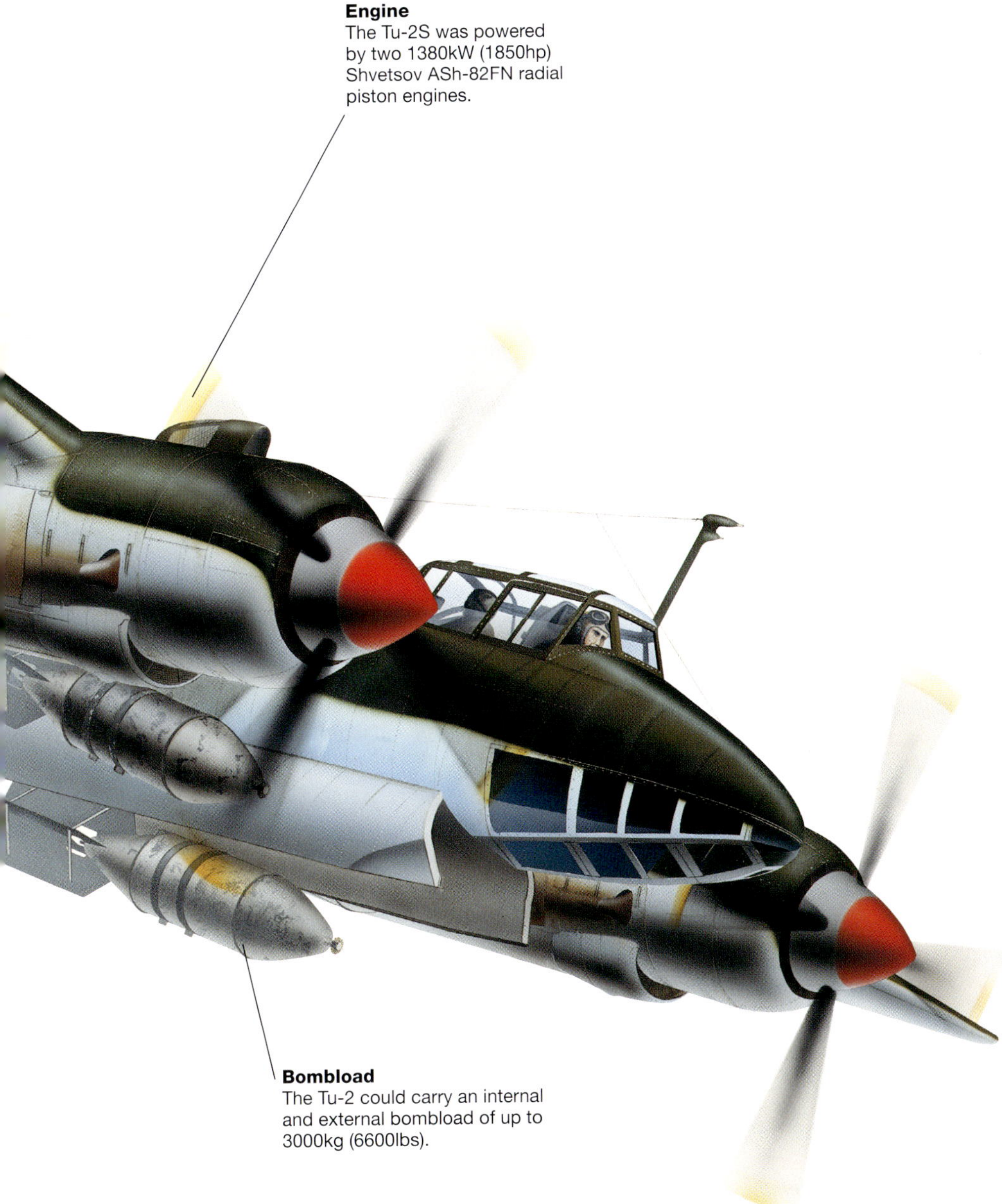
Engine
The Tu-2S was powered by two 1380kW (1850hp) Shvetsov ASh-82FN radial piston engines.
Bombload
The Tu-2 could carry an internal and external bombload of up to 3000kg (6600lbs).

Douglas A-20

More Douglas A-20s served the Soviet Union during World War II than any other type of Lend-Lease aircraft. Fast, easy to fly and reliable, the A-20 was popular and remained in Soviet service to the end of hostilities and beyond.

Douglas A-20B

Serving with the Black Sea Feet Air Force in 1944, this A-20B has been fitted with a Soviet-made UTK-1 turret. The emblem on the nose depicts a wild goose flying over the Black Sea.

Douglas A-20B
Weight (Maximum take-off) 10,795kg (23,800lbs)
Dimensions Length: 14.63m (48ft 0in), Wingspan: 18.69m (61ft 4in), Height: 5.51m (18ft 1in)
Powerplant Two 1200kW (1600hp) Wright R-2600-11 Double Cyclone air-cooled radial piston engines
Speed 563km/h (350mph)
Range 1521km (945 miles)
Ceiling 8717m (28,600ft)
Crew 3
Armament Two 12.7mm (0.5in) Browning M2 machine guns fixed forward-firing in the nose, one 12.7mm (0.5in) Browning M2 machine gun flexibly mounted in dorsal position, one 7.62mm (0.3in) machine gun flexibly mounted in ventral tunnel position; up to 1089kg (2400lbs) bombload

The Douglas Model 7B first flew on 23 January 1939, entering service with the USAAF in January 1941. By that time the type had seen extensive combat service with the French as the DB-7 and with the British as the Douglas Boston. When the Soviet Union entered the war in June 1941, 277 Douglas Bostons, drawn from RAF stocks, were quickly shipped to Murmansk. This probably explains why the aircraft was generally referred to as 'Boston' throughout its Soviet service. Further deliveries came direct from the Douglas factory in the US, the majority of these being flown to their destination, either via Iran or westwards via Alaska. Ultimately 3414 A-20s were delivered to the Soviet Union, making it the largest user of the type.

Prolific service

First combat use of the A-20 in Soviet hands occurred during 1942 and the Bostons would operate effectively over every major battle from Stalingrad through to the final assault on Berlin. Although the VVS crews appreciated its speed and easy handling, they were less pleased with the light armour and defensive armament of four 7.62mm (.30in) Browning machine-guns. These weapons were replaced in Soviet service by the ShKAS machine gun of the same calibre but with much greater rate of fire, sometimes fitted in Russian-built turrets, the first of many modifications that the aircraft would undergo in Soviet service.

Armour protection was increased, although the aircraft was considered unsuitable for its intended attack role due to its vulnerability and was mainly employed as a medium-altitude bomber. In addition, modern Soviet bombsights, either the NKPB-4 or OPB-1, replaced the obsolescent Whimpers D-8 bombsight the aircraft was delivered with, the Soviets being extremely disappointed that the A-20s were not supplied with the much vaunted Norden bombsight. The subtype received in the largest numbers was the A-20G, of which 1530 were supplied, and this was a dedicated attack variant with a 'solid' nose featuring four 20mm (0.78in) Hispano cannon and two 12.7mm (0.5in) M2 Browning machine guns replacing the bombardier's glazed nose of earlier models. As the Soviets were loath to employ the A-20 in the attack role, a great deal of effort was spent modifying the A-20G to carry a navigator/bomb-aimer. Various modifications were tried but the definitive solution was to remove the cannon and fit a seat for a navigator in the nose with relatively minimal glazing. The considerable amount of weight lost from the nose was balanced by adding more armour plate. It is unknown how many A-20Gs were converted, probably several hundred. At least one modified aircraft was allocated per squadron to act as a 'lead ship' in bombing operations.

Anti-shipping role

As well as its major employment as a bomber, the A-20 was also utilized with considerable success by naval units as an anti-shipping aircraft, carrying one or two 45-36 AN torpedoes externally or 500kg (1102lb) or 1000kg (2204lb) mines. A-20s were part of an August 1944 raid on the Romanian port of Constanta that effectively ended Axis operations on the Black Sea and in May 1945, A-20Gs sank the 15,241 tonne (15,000 ton) former commerce raider *Orion*. A Naval A-20 was also the first Soviet aircraft to reconnoitre the Norwegian port of Narvik, the aircraft being widely employed as a long-range reconnaissance platform. Indeed, its final wartime use was for reconnaissance operations over Hokkaido in August 1945. Around 900 A-20s remained in Soviet service at VJ day and the aircraft received the NATO reporting name 'Box'.

ALLIED STRATEGIC BOMBERS

Throughout the war the Soviet Union made repeated requests for long-range Western strategic bombers to be supplied via Lend-Lease. For political reasons virtually none was ever supplied, the singular exceptions being a B-24 that had ended up stranded in Yakutsk and was handed over to Soviet authorities, and a British Short Stirling that was supplied for testing during 1944, by which time the type was obsolescent.

Nonetheless the USSR determined to gain examples of modern Western long-range bombers and Soviet authorities undertook to rebuild as many airworthy examples as possible, their attention first landing on six damaged Avro Lancasters abandoned near Arkhangelsk in the north of Western Russia. These aircraft originally belonged to 617 Squadron (the famed 'Dambusters') and had operated from the USSR during September 1944 to bomb the *Tirpitz*. Soviet engineers selected the two least damaged examples, stripped them of armament and returned these two to airworthy condition, although the second only flew after the conclusion of hostilities.

The first of the Lancasters, coded 01, was used intensively from January 1945 for escorting convoys, ice reconnaissance and maritime patrol, tasks for which its great range rendered it ideal. Both Lancasters operated for a time after the war – '01' was sent to a technical school in Riga and its ultimate fate is unknown; the second example was written off in a landing accident in 1946.

In addition to the Lancaster, the Soviets made a Herculean effort to place the Consolidated B-24 Liberator into service. Some 30 airworthy examples of the Liberator were rebuilt from 73 force-landed examples on Soviet territory, but the first only entered service after the end of hostilities. The B-24s served until at least 1949, latterly as a conversion trainer for crews transitioning to the Tupolev Tu-4 (a reverse-engineered B-29 Superfortress).

Consolidated B-24H Liberator

Weight (Maximum take-off) 32,295kg (71,200lbs)

Dimensions 20.47m (67ft 2in), Wingspan: 20.47m (110ft 0in), Height: 5.49m (18ft)

Powerplant Four 890kW (1,200hp) Pratt & Whitney R-1830 Twin Wasp, 14-cylinder two-row air-cooled turbosupercharged radial piston engines

Speed 478km/h (297mph)

Range 2840km (1540 miles)

Ceiling 8500m (28,000ft)

Crew 11

Armament Two 12.7mm (0.5in) M2 Browning machine guns in nose turret, dorsal turret, tail turret and retractable ventral turret, one 12.7mm (0.5in) M2 Browning machine gun flexibly mounted in each waist position; up to 3600kg (8000lbs) bombload

Consolidated B-24H Liberator

'Red 7' was a B-24H delivered to 45 TBAD in July 1945, but crews were still in training when Japan surrendered. Retaining its US serial on the fin, this Liberator was one of three serving with 890 BAP on Tu-4 transition training from 1948 onwards.

Handley-Page Hampden

The Handley-Page Hampden was approaching the end of its career when, somewhat unexpectedly, 17 examples were operated by the Soviets in the anti-shipping role.

After the mauling of Convoy PQ-17 on its journey with war material to the Soviet Union during July 1942, urgent means were sought to protect future convoys venturing through the Arctic Sea. To this end Operation Orator saw 144 and 455 squadrons of RAF Coastal Command fly their Handley-Page Hampden TB.1 torpedo bombers to Vaenga, near Murmansk, to supply anti-shipping cover to the next convoy, PQ-18, on its approach to the USSR. Although the RAF crews flew a limited amount of sorties, the mere presence of the Hampdens deterred German surface raiders from attacking.

Gift

The operation was declared a success and the aircraft prepared to leave for the UK. However prevailing westerly winds made the return journey potentially hazardous and the decision was made instead to give the 17 surviving aircraft to the Soviet Navy, a proposal agreed by Churchill on 6 October 1942. Operated by the 3rd squadron of the 24th Mine-Torpedo Aviation Regiment (MTAP) of the Northern Fleet, the Hampden's unusual shape led to its Russian nickname of '*Balalaika*' and the aircraft were used intensively until the end of 1943 when the spares supply had been exhausted.

Hero of the Soviet Union

Usually operating in pairs, the Hampdens proved successful in the anti-shipping role, sinking several vessels, their largest single victory being the sinking of a 12,192 tonne (12,000 ton) transport on 29 January 1943. In one action Captain A.A. Bashtyrkov's Hampden was set on fire but he pressed home his torpedo attack, managing to sink an 8128 tonne (8000 ton) transport before crashing in flames into the sea. Bashtyrkov was posthumously awarded Hero of the Soviet Union for this action.

Handley-Page Hampden

Formerly operated by 455 Squadron RAAF (Royal Australian Air Force) and wearing standard RAF night-bomber camouflage with painted-out roundels and squadron codes, Hampden TB.1 'White 30' was based at Vaenga near Murmansk in 1942.

Handley-Page Hampden

Weight (Maximum take-off) 10,206kg (22,500lbs)

Dimensions Length: 16.33m (53ft 7in), Wingspan: 21.08m (69ft 2in), Height: 4.55m (14ft 11in)

Powerplant Two 750kW (1000hp) Bristol Pegasus XVIII 9-cylinder air-cooled radial piston engines

Speed 398km/h (247mph)

Range 2770km (1720 miles)

Ceiling 5800m (19,000ft)

Crew 4

Armament One fixed forward-firing 7.7mm (0.303in) Browning machine gun in nose, one 7.7mm (0.303in) Vickers K machine gun flexibly mounted in nose, one or two 7.7mm (0.303in) Vickers K machine guns flexibly mounted in dorsal mount, one or two 7.7mm (0.303in) Vickers K machine guns flexibly mounted in ventral mount; up to 1800kg (4000lbs) bombload or one 457mm (18in) torpedo or mines

North American B-25

In total, 862 examples of the versatile North American B-25 served in the Soviet Union and this excellent medium bomber saw combat in Russian hands from the Battle of Stalingrad onwards.

North American B-25D

Piloted by Captain Kavkaev of 229 GvBAP in January 1945, this B-25D has the name '*Mstitel*' ('Avenger') painted on the nose.

The first B-25 to enter Soviet service was a single example from the famed Doolittle raid on Tokyo of April 1942. Launched from the carrier USS *Hornet*, the B-25B was interned by Soviet authorities when it landed at Vozdvizhenka airfield near Vladivostok and was maintained in airworthy condition until it was accidentally destroyed in a hangar fire in 1950.

Robust service record

All other B-25s arrived by more conventional means and the aircraft was popular with crews being reliable, well-armed and easy to fly. In action from late 1942 onwards, it became quickly apparent that the typical VVS low-level, close support bombing operations did not utilize the aircraft's strengths and the B-25's excellent range, good speed, powerful defensive armament and decent bombload (all of which were superior to the domestically produced Il-4), saw it used primarily by the ADD for long-range operations against the German rear from September 1942 until VE day. The only criticism of any note was the lack of a tail gun and the aircraft's relative vulnerability to stern attack. Field modifications added a single rear gun to the earlier models, later B-25Js featured a factory fitted tail turret with twin 7.62mm (0.3in) Browning machine-guns.

Transport modification

During the course of a busy career with Soviet forces B-25s bombed targets including Warsaw, Königsberg and Berlin. Some were modified into transport aircraft, capable of carrying up to 20 (cramped) passengers or a tonne of cargo. Under the terms of the Lend-Lease agreement all B-25s were supposed to be destroyed or returned, but although most were indeed scrapped, several units retained the B-25 and it remained in Soviet service until at least 1949, NATO assigning it the reporting name 'Bank'.

North American B-25D

Weight (Maximum take-off) 18,960kg (41,800lbs)

Dimensions Length: 16.10m (52ft 11in), Wingspan: 20.60m (67ft 7in), Height: 4.80m (15ft 9in)

Powerplant Two 1300kW (1700hp) Wright R-2600-92 Twin Cyclone 14-cylinder two-row air-cooled radial piston engines

Speed 443km/h (275mph)

Range 2414km (1500 miles)

Ceiling 7315m (24,000ft)

Crew 6

Armament One 12.7mm (0.5in) Browning M2 machine gun flexibly mounted in nose, two 12.7mm (0.5in) Browning M2 machine guns in dorsal turret, two 12.7mm (0.5in) Browning M2 machine guns flexibly mounted in port and starboard waist positions, two 12.7mm (0.5in) Browning M2 machine guns flexibly mounted in tail position; up to 1814kg (4000lbs) bombload

RECONNAISSANCE AND TRANSPORT AIRCRAFT

The transport capability of the VVS was primarily US designed, in the form of the Douglas C-47 or its licence-built counterpart, the Lisunov Li-2, though obsolete bombers and a plethora of small utility types also served. Gliders often provided transport to partisan groups. Few reconnaissance types were fielded, the VVS tending to use modified bombers for this task.

This chapter includes the following aircraft:

- Tupolev R-6
- Polikarpov R-5
- Polikarpov R-Z
- Kharkov R-10
- Tupolev ANT-9
- Vultee V-11
- Lisunov Li-2
- Kamov A-7
- Gribovsky G-11
- Kolesnikov-Tsybin KC-20
- Antonov A-7
- Yakovlev Yak-6
- Shcherbakov Shche-2
- Curtiss O-52 Owl
- Douglas C-47
- Armstrong Whitworth Albemarle

A Lisunov Li-2, circa 1943. A licence-built version of the American Douglas DC-3, the Li-2 was used for transport, partisan supply, bombing and as ambulance aircraft from 1942 onwards.

Tupolev R-6

Intended as a multi-role aircraft, the Tupolev R-6 was long past its prime by 1941 but provided useful service in second line roles.

First flying in September 1929, Tupolev's ANT-7 was essentially a scaled down Tupolev TB-1. The aircraft was intended to function not only as a reconnaissance aircraft and medium bomber but also be able to engage in aerial combat as a heavy fighter. Tupolev's design was an all-metal twin engine cantilever monoplane that was very advanced for its era.

Armament

Operated by a crew of four, it could carry 500kg (1102lbs) of bombs externally and was armed with five British 7.7mm (0.303in) Lewis machine guns, two in the nose and dorsal positions, with a further Lewis in the ventral position. Unusually for the era, standard equipment included a radio.

Flight testing revealed the aircraft, powered by a pair of imported BMW VI engines, was only marginally slower than contemporary fighters and it was ordered into production in 1931 as the R-6.

Production aircraft were fitted with domestically produced M-17 engines of slightly lower power, and as such performance was marginally lower although still considered to be well up to international standards for this type of aircraft.

Naval variant

A naval variant, the MR-6 floatplane, was developed for the torpedo bombing role, and civilian variants were operated by Aeroflot and Avia Arktika. A civil R-6 gained brief fame for the type when it became the first Soviet aircraft to fly over the North Pole on 5 May 1937. Civil floatplanes were designated MP-6 and were widely used, mainly in the North and

Tupolev R-6

Weight (Maximum take-off) 6472kg (14,268lbs)
Dimensions Length: 14.75m (48ft 5in), Wingspan: 23.2m (76ft 1in), Height: 4.95m (16ft 3in)
Powerplant Two 540kW (720hp) Mikulin M-17F V-12 water-cooled piston engines
Speed 230km/h (140mph)
Range 800km (500 miles)
Ceiling 5620m (18,440ft)
Crew 4
Armament Two 7.7mm (0.303in) Lewis machine guns flexibly mounted in nose, one 7.7mm (0.303in) Lewis machine gun in ventral 'dustbin' turret, two 7.7mm (0.303in) Lewis machine guns flexibly mounted in dorsal position; up to 500kg (1102lbs) bombload

Tupolev R-6

'Yellow 14' was operating in an unknown unit during the mid 1930s. Tupolev had close ties with Junkers and the R-6 utilized a corrugated duralumin skin of great strength originally developed by the German company.

Siberian River Airlines, until 1945. Thoroughly obsolete by the outbreak of war, the remaining R-6s were operated mainly as transport aircraft, being used to transfer flight personnel when redeploying between airfields, supplying units with ammunition, spare parts and fuel, as a communications aircraft and for transporting medicines and the wounded. Despite their advanced years, these dependable aircraft worked hard throughout the Great Patriotic War.

Tupolev MR-6
Weight (Maximum take-off) 6410kg (14,132lbs)
Dimensions Length: 16m (52ft 6in), Wingspan: 23.2m (76ft 1in), Height: 5.45m (17ft 11in)
Powerplant Two 507kW (680hp) Mikulin M-17 V-12 water-cooled piston engines
Speed 235km/h (146mph)
Range 1700km (1056 miles)
Ceiling 5000m (18,440ft)
Crew 2–4
Armament One 7.7mm (0.303in) Lewis machine gun flexibly mounted in nose, two 7.7mm (0.303in) Lewis machine guns flexibly mounted in dorsal position; one 450mm (18in) TAN-12 torpedo

Tupolev MR-6
In floatplane form as the MR-6, the aircraft dispensed with the ventral gun position to allow for a torpedo to be carried. This completely anonymous example wears the typical naval pre-war scheme of green over light grey.

Polikarpov R-5

The Polikarpov R-5 was the standard Soviet reconnaissance and light bomber aircraft. It was built in huge quantities for its era and served well into World War II.

Designed by Nikolai Polikarpov as a replacement for the R-1, itself an unlicensed copy of the British Airco DH.9A, the R-5 was a thoroughly conventional single engine, two seat, single bay biplane.

Powered by a 507kW (680hp) Mikulin M-17 engine, a licence-built BMW VI, the R-5 was first flown during the autumn of 1928 and it entered service with the VVS in 1930. Before its service entry the new aircraft made a non-stop flight from Moscow to Sevastopol, piloted by Viktor Pisarenko and Deputy Chief of the Red Army Air Force Yakov Alksnis. Alksnis was obviously impressed with the Polikarpov as he kept a snow-white R-5 as his personal aircraft until his arrest in 1937.

In service the R-5 proved dependable and popular. It served in action throughout the mid to late 1930s in the Spanish Civil War, over Khalkhin Gol, with Soviet volunteer units in the Sino-Japanese war, both in standard light bomber form and also as the R-5Sh, the *Shturmovik* version with four PV-1 7.62mm (0.3in) machine guns mounted in two pods under the lower wing supplementing the standard armament of one forward firing fixed PV-1, with a second such weapon flexibly mounted in the observer's cockpit for rear defence.

An early Polikarpov R-5 is manhandled into position. The R-5 proved popular with crews during the Winter War in Finland, but was obselete by the time of the German invasion in June 1941.

Winter War service

Heavily committed during the Winter War with Finland of 1939–40, the R-5 was obsolete by 1941. Nevertheless it performed useful work, generally as a nocturnal harassment bomber against the Axis invaders, remaining in front line service until at least early 1944.

In addition to its combat utilization the R-5 was widely employed as a liaison aircraft, trainer and light transport, around 1000 of the approximately 7000 built being of a civil version for Aeroflot, late examples possessing an enclosed cabin for four passengers. Other variants of this ruggedly capable aircraft included a torpedo bomber version and a floatplane of which 111 were built.

Polikarpov R-5

Weight (Maximum take-off) 3247kg (7158lbs)
Dimensions Length: 10.56m (34ft 8in), Wingspan: 15.5 m (50ft 10in), Height: 3.25m (10ft 8in)
Powerplant One 507kW (680hp) Mikulin M-17B V-12 liquid-cooled piston engine
Speed 228km/h (142mph)
Range 800km (500 miles)
Ceiling 6400m (21,000ft)
Crew 2
Armament One 7.62mm (0.3in) PV-1 machine gun fixed forward firing in upper cowling, one 7.62mm (0.3in) Degtyaryov DA machine gun flexibly mounted in rear cockpit

Polikarpov R-5SSS

SSS stood for *Skorostnoy, Skoropod'yemnyy, Skorostrel'nyy* or 'High speed, Rapid climb, Rapid fire'. 'Yellow 12' was operating with an unknown unit in the Kiev Military District during 1936.

Polikarpov R-Z

A more powerful and better streamlined development of the successful R-5, the Polikarpov R-Z delivered a useful improvement in performance at the cost of ease of handling.

First flown in January 1935 and in service by the end of the year, the R-Z was derived from the R-5-SSS, a variant of the R-5 with improved aerodynamics and heavier armament of which around 100 were built. The R-Z took this process further and combined an aerodynamically cleaned up airframe with a more powerful engine, the 600kW (805hp) Mikulin AM-34, pushing the maximum speed up to a creditable 316km/h (196mph) at 3500m (11,483ft) and allowing an increase in bombload to 400kg (882lbs). The fixed, forward firing gun armament remained a single 7.62mm (0.3in) PV-1 but the rear weapon was often one, or sometimes two, of the new ShKAS rapid fire machine gun of the same calibre but that delivered a much improved weight of fire.

Supplemental role

Produced until the spring of 1937, a total of 1031 R-Zs was built and saw widespread action, similar to the R-5, in Spain and the Far East before being employed in the invasion of Poland and the Winter War against Finland of 1939. Throughout its career the R-Z supplemented rather than replaced the R-5, a situation that – somewhat counterintuitively – became more pronounced during World War II.

Like the R-5, it was quickly apparent that the R-Z was extremely vulnerable when operated by day, however it was not an easy aircraft to fly and, although its range and relatively heavy armament were appreciated, its tricky handling – especially on takeoff and landing – was far from ideal for operation by night.

Battle of Moscow

By the time of the Battle of Moscow in November 1941, five regiments were operational with the R-Z compared to 27 equipped with the older R-5 and by the end of 1942 the R-Z's fighting career was effectively over.

Polikarpov R-Z

Weight (Maximum take-off) 3150kg (6945lbs)
Dimensions Length: 9.72m (31ft 11in), Wingspan: 15.45m (50ft 8in), Height: 3.6m (11ft 10in)
Powerplant One 600kW (805hp) Mikulin AM-34 V-12 liquid-cooled piston engine
Speed 316km/h (196mph)
Range 1000km (620 miles)
Ceiling 8700m (28,500ft)
Crew 2
Armament One 7.62mm (0.3in) PV-1 machine gun fixed forward firing in upper cowling, one or two 7.62mm (0.3in) ShKAS machine guns flexibly mounted in rear cockpit

Polikarpov R-Z

The R-Z had proved successful over Spain but by the autumn of 1941, when this example was in service with 719 LBAP, the aircraft was painfully vulnerable to enemy fighters, necessitating a switch to nocturnal operations.

Kharkov R-10

Developed from Europe's first cantilever monoplane passenger aircraft with retractable undercarriage, production of the promising R-10 design was cut short by the arrest of its designer Joseph Neman.

Intended primarily for reconnaissance with a secondary light attack role, the aircraft then known as the Kharkov Aviation Institute KhAI-5, first flew in June 1936. Performance was very good. Powered by the 531kW (712hp) Shvetsov M-25, a maximum speed of 388km/h (241mph) was achieved and the aircraft proved easy to fly – test pilots were full of praise for the aircraft. It was recommended that the KhAI-5 become the primary reconnaissance aircraft for 1937. Re-designated the R-10, it entered production in July 1937 and the first examples were in squadron service by the end of the year.

Stressed shell structure

Of wooden semi-monocoque construction, the two-seat R-10 was armed with two fixed forward firing ShKAS 7.62mm (0.3in) machine guns in the upper nose decking with a further ShKAS mounted in an MV-3 turret for rear defence. The bombload was a modest 300kg (661lbs) carried in an internal bomb bay mounted between the two crew positions. An APA-10 camera was fitted in the floor of the rear cockpit for reconnaissance duties.

Kharkov R-10

Pictured as it appeared during with 9 LBAP during 1939, this R-10 was flown by the commanding officer of 5 AE, Major B. Uspenskiy.

Night missions

The R-10 was committed to combat in the invasion of Poland in 1939 and the subsequent Winter War against Finland, but by the time of the German invasion it was becoming outdated. Many were destroyed on the ground in the opening phase of the conflict but enough escaped to participate in the battles for Leningrad and Moscow flying reconnaissance and close support missions. The vulnerability of the aircraft to fighters forced missions to increasingly be flown by night, combat usage continuing into 1943 when all surviving R-10s were withdrawn.

Neman was working on the KhAI-52, which promised to be a considerably more formidable aircraft, when he was arrested on trumped-up charges of espionage and his design bureau was closed down. Unlike many of his contemporaries he did at least survive his incarceration but work on the KhAI-52 was abandoned.

Kharkov R-10

Weight (Maximum take-off) 2515kg (5545lbs)
Dimensions Length: 9.3m (30ft 6in), Wingspan: 12.2m (40ft), Height: 3.80m (12ft 6in)
Powerplant 531kW (712hp) Shvetsov M-25 9-cylinder air-cooled radial piston engine
Speed 388km/h (241mph)
Range 1450km (900 miles)
Ceiling 7700m (25,300ft)
Crew 2
Armament Two 7.62mm (0.3in) ShKAS machine-guns fixed forward firing in cowling, one 7.62mm (0.3in) ShKAS machine-gun in rear turret; up to 400kg (882lbs) bombload

Tupolev ANT-9

The ANT-9 was the most produced large Soviet airliner of the inter-war period and saw long service as both a civil and military transport.

Like its more famous contemporary the Junkers Ju 52/3m, the ANT-9 was designed to be primarily an airliner but with the secondary ability to operate as a bomber. It used existing components to speed the development process, the wings and tail for example being taken from the Tupolev R-6. Despite the first prototype suffering a collision with a tram while being removed from display in Red Square, the ANT-9 was first flown on 5 May 1929 powered by three Gnome Rhone Titans and after successful testing entered production in early 1930. Production aircraft were fitted with Wright J-5s, but although these aircraft proved reliable they were expensive and the Bessonov M-26 was used instead. Unfortunately that engine proved extremely unreliable and was abandoned by 1933, a switch being made to two M-17 engines.

Deruluft route

Most civilian ANT-9s operated from Moscow into the far east of Russia but the aircraft was also used on the Deruluft route to Berlin and flew to other European destinations. From 1935 the aircraft was known as the PS-9 and the last were withdrawn from Aeroflot use in early 1941. By this time the aircraft had seen active military service, two examples having operated as medical transports during the 1940 Winter War against Finland.

None were lost when Germany launched its attack as all were based far from the Western border, and subsequently the aircraft was generally employed well behind the front line. One exception was a pair of PS-9s of the 8th Transport Regiment used for the nocturnal delivery of reconnaissance and sabotage groups behind enemy lines. Making up to six sorties per night, these aircraft transported 655 people and 68 tonnes (67 tons) of cargo.

Of 16 Ps-9s listed operational in September 1941, 10 were still serviceable in December 1943 but all had disappeared by August 1944.

Tupolev ANT-9

Weight (Maximum take-off) 5043kg (11,118lbs)

Dimensions Length: 16.8m (55ft 1in), Wingspan: 23.7m (78ft 9in), Height: 4.86m (15ft 11in)

Powerplant Three 172kW (230hp) Gnome-Rhône 5K Titan 5-cylinder air-cooled radial piston engines

Speed 215km/h (134mph)

Range 1800km (1118 miles)

Ceiling 3750m (12,300ft)

Crew 2

Armament None

Tupolev ANT-9

The all-metal ANT-9 was a decent aircraft that initially suffered from the lack of a good engine. Early examples such as this were trimotors but by the time the Great Patriotic War broke out all had been re-engined with two Mikulin M-17s.

Vultee V-11

The American Vultee V-11 was designed as a light attack aircraft but all Soviet examples were used in courier and light transport roles.

In the early 1930s Vultee developed an advanced transport aircraft, the V-1A, which on its introduction in 1934 was briefly the fastest airliner in the world. Unfortunately for Vultee, new US regulations on single engine airliners crippled sales and only 25 were built. Seeking alternative markets for the design, Vultee utilized a redesigned fuselage to develop the V-11 light attack aircraft. Development was rapid, the first example taking to the air on 17 September 1935.

Armoured attack aircraft

During 1936, Soviet trade representatives obtained a manufacturing licence for the V-11GB and four examples to serve as pattern aircraft for production in Russia. In 1937 the first Soviet built V-11 was completed, designated BSh-1 for *Bronirowanni Shturmovik* – 'Armoured attack aircraft' – the Russian-built aircraft featured an M-62 engine and, as the designation suggests, had been fitted with armour, the original V-11 featuring no such protection.

Light passenger duties

Unfortunately the addition of armour and military equipment increased the weight and degraded the performance of the BSh-1 so much that it was only 42km/h (26mph) faster than the biplane Polikarpov R-Z and could be out-climbed by the older aircraft. After failing its state acceptance tests production of the BSh-1 was stopped in 1938 after 36 had been built, the first five from US manufactured components.

Stripped of military equipment, the aircraft were assigned to Aeroflot in 1939 for conversion to airmail and light passenger duties as a fast three-passenger airliner designated PS-43 and were used extensively.

During the war, PS-43s flew mail and courier flights and some were armed with a defensive machine gun after several were shot down by German fighters. The aircraft was reliable and quite effective in service, and three were still operational transporting mail to the Ukrainian and Belorussian fronts as late as January 1945.

Vultee V-11
Weight (Maximum take-off) 7387kg (16,285lbs)
Dimensions Length: 11.53m(37ft 10in), Wingspan: 15.24m (50ft), Height: 3.05m (10ft)
Powerplant One 890kW (1200hp) Pratt & Whitney R-1830-17 Twin Wasp 14-cylinder air-cooled radial engine
Speed 370km/h (230mph)
Range 1790km (1110 miles)
Ceiling 6200m (20,500ft)
Crew 1
Armament One 7.62mm (0.3in) ShKAS machine gun flexibly mounted in rear cockpit on some examples

Vultee V-11

With Aeroflot emblazoned on the nose, this V-11 was undergoing flight testing at NII GVF, the Research Institute of Civil Aviation, in 1939.

Lisunov Li-2

A licence-built Douglas DC-3, over 4000 Li-2s were produced and the aircraft saw service both as a transport and long-range bomber.

Originally operated by Aeroflot from 1939 as the PS-84, the aircraft was not a straight copy of the US aircraft and was powered by two Shvetsov M-62 radials of lower power than the Pratt & Whitney R-1830s fitted to American aircraft. The engines also benefitted from a faceplate with adjustable shutters to deal with cold conditions. Heavier than its progenitor due to the closest equivalent metric aluminium for the skin being slightly thicker than its imperial equivalent, the PS-84 also had a slightly shorter wingspan.

By the outbreak of war in 1941, 237 PS-84s had been built but production paused while the factory was evacuated to Tashkent in Uzbekistan, restarting in January 1942. Later in the

Lisunov Li-2

Sporting the standard three-colour scheme worn by most late-war Li-2s, this example was operated by 12 GvAP DD at Helsinki during the winter of 1944.

Lisunov Li-2

Flying from Dubrovka airfield in central Ukraine in 1944, this Lisunov has black-painted undersides. This finish was often applied to bomber variants of the Li-2, though by 1944 most had reverted to their original transport role.

Lisunov Li-2
Weight (Maximum take-off) 11,280kg (24,867lbs)
Dimensions Length: 19.65m (64ft 6in), Wingspan: 28.81m (94ft 6in)
Powerplant Two 670kW (900hp) Shvetsov M-62 9-cylinder air-cooled radial piston engines
Speed 280km/h (170mph)
Range 2500km (1600 miles)
Ceiling 5600m (18,400ft)
Crew 4–6
Armament One 12.7mm (0.5in) Berezin UBT machine gun in dorsal turret, two 7.62mm (0.3in) ShKAS machine guns flexibly mounted in port and starboard side hatches; up to 2000kg (4409lbs) bombload

In military service, the Li-2 was equipped with a 12.7mm (0.5in) UBT heavy machine gun in a dorsal turret, as shown in this photograph.

same year the aircraft was redesigned as the Li-2 in accordance with the new Soviet aircraft designation system.

Battle of Stalingrad

Many PS-84s were impressed into military service and thousands more Li-2s were built for VVS use. Unlike US C-47s, Li-2s were often armed with a dorsal UTK-1 turret mounting a single 12.7mm (0.5in) UBT machine gun and provision was made for firing a 7.62mm ShKAS from a beam position on either side of the aircraft.

The most produced Soviet transport aircraft of the war, it remained in service for decades after the conflict. The Li-2 was also widely used as a bomber, which were initially conversions of transport aircraft but later became the Li-2NB, a purpose built bomber version.

Despite lacking a bomb bay, the Li-2 could carry a maximum bombload of 2000kg (4409lb) under the centre section and was fitted with extra defensive machine guns. Mostly used to re-equip units operating the obsolete TB-3, the Li-2NB participated in long-range bombing from the 1942 Battle of Stalingrad through to the end of hostilities.

A comfortable aircraft due to its airliner origins, featuring both a galley and a toilet, the Li-2NB was very popular with crews.

Kamov A-7

The world's first rotary winged aircraft to see combat service, only five Kamov A-7 autogyros were built and saw brief action during 1941.

The two-seat A-7 that had flown as early as 1934 was subject to unusually thorough testing, representing as it did the first example of a completely new type of aircraft for the Soviet Union. A second prototype took to the air in 1937 and the decision was eventually made to manufacture five production autogyros in 1940. Designed by Nikolai Kamov, the two-seat A-7 was intended to function as a tactical reconnaissance and artillery spotting aircraft.

'Aviation Day' display

Both Britain and France operated autogyros during the war's early stages but these were not armed. However the A-7 featured a fixed forward firing 7.62mm (0.3in) PV-1 machine gun and a pair of 7.62mm Degtyarev DA for rear defence, as well as a bombload of up to 400kg (882lbs) or six RS-82 rockets. During testing problems were encountered with vibration of the rotor and tail and engine overheating but these were overcome and the A-7 was demonstrated to the public on 'Aviation Day', 18 August 1935.

During the spring of 1941 the A-7 was successfully employed as a crop duster to deal with an outbreak of apple moth on the mountainside orchards in Tien Shan, Kyrgyzstan. It was found during this operation that not only could the A-7 operate from small landing site but that it could also fly in valleys and down steep mountain slopes that would be inaccessible to conventional aircraft. Mere months later the A-7 would be fighting the Wehrmacht rather than insects. Five A-7s were organized into an autogyro squadron and flew 19 combat sorties by day and night correcting artillery fire. Although none of the A-7s was lost to enemy action, the final flight by a Soviet autogyro was made on 5 October 1941 by which time all the A-7s were unserviceable. Kamov was planning an improved version but the chaotic conditions prevailing after the German invasion prevented further autogyro development.

Kamov A-7
Weight (Maximum take-off) 2390kg (5270lbs)
Dimensions Length: 8.4m (27ft 6in), Wingspan: 10.40m (34ft 2in), Height: 3.88m (12ft 8in)
Powerplant One 358kW (480hp) Shvetsov M-22 9-cylinder air-cooled radial piston engine
Speed 218km/h (135mph)
Range 400km (250 miles)
Ceiling 4760m (15,617ft)
Crew 2
Armament One 7.62mm (0.3in) PV-1 machine gun fixed forward-firing in cowling, two 7.62mm (0.3in) Degtyarev DA machine guns flexibly mounted in rear cockpit; up to 400kg (882lbs) bombload or six RS-82 rockets

Kamov A-7
Shown as it appeared when attached to the Soviet 24th Army during the Yelnya offensive of August–September 1941, Kamov's A-7 was the world's first combat rotorcraft.

Gribovsky G-11

The most produced Soviet transport glider of the war, the Gribovsky G-11 was heavily used from 1942 onwards.

On 7 July 1941 following the outbreak of hostilities, glider designer Vladimir Gribovsky was asked how long it would take to plan and build a new 11-seat assault glider. Gribovsky maintained he would need two months, and remarkably the first example of his G-29 took to the air on 1 September. Full scale production was approved within three weeks and the first production gliders began to appear in early November. Designated G-11, due to the 11 paratroopers they carried, production continued until early June 1942 when the decision was made to cease after 308 had been built due to a shortage of tug aircraft and glider pilots. There were concerns that the G-11s airframe would not stand up well to prolonged open-air storage and it made more sense to build more when the requirement arose.

Reinforced structure

Production was indeed resumed at the end of 1943 and an improved version was developed featuring a reinforced structure, this version remaining in production until 1948. From October 1944 a training version was developed as the G-11U, which featured dual controls. A powered G-11 was also built and tested, featuring a pylon-mounted engine above the fuselage but this did not enter production.

Partisan operations

Similar to the A-7, the G-11 was heavily employed supplying partisan groups fighting behind enemy lines and was regarded as the superior of the two due to its greater load-carrying capacity. The most intense period of operations occurred from March to November 1943, when hundreds were used to transport supplies to partisans fighting as part of the Kalinin Front in Belarus. Gliders were flown in by night and burned once their loads had been delivered. The sole known exception was a G-11 that was used to ferry out two wounded partisan commanders in April 1943, towed by a Tupolev SB.

Gribovsky G-11

Regarded as essentially disposable, Soviet transport gliders seldom had any sort of individual markings, and this G-11 is typical. The undercarriage was fixed but could be folded upward, allowing the aircraft to land on its central skid, shortening the landing run.

Gribovsky G-11

Weight (Maximum take-off) 2400kg (5291lbs)

Dimensions Length: 9.8m (32ft 2in), Wingspan: 18m (59ft 1in), Height: 2.7m (8ft 10in)

Speed 280km/h (170mph)

Crew 1 plus 10 troops

Armament None

Kolesnikov-Tsybin KC-20

The KC-20 was the least produced of the operational Soviet gliders but it could carry the greatest payload.

Designed in the summer of 1941, the KC-20 possessed an excellent load capacity, being able to carry up to 20 equipped troops or up to 2000kg (4409lbs) of cargo. Two prototypes were completed in October 1941 and the aircraft was ordered into production following successful flight tests, although ultimately only 68 examples of the KC-20 were built.

As well as being considerably larger than its contemporaries, the KC-20 was distinctive for its twin fin and rudder tailplane, this being a legacy of the original design specification that envisioned the fitting of a gun position in the tail for defence.

The armament requirement was dropped but the twin tail remained. In service the KC-20 was used in a similar fashion to the smaller A-7 and G-11, in the main resupplying partisan groups, although in November the KC-20 was used to urgently ferry large quantities of antifreeze to tank units operating in the defence of Stalingrad.

Cargo disassembly

Despite its impressive lifting capacity, the KC-20 featured a relatively small loading door that led to problems in service; for example, relatively large artillery pieces were able to be carried by the glider but had to be dismantled in order to be loaded and unloaded. The placement cockpit position at the extreme nose of the aircraft was also problematic as this could lead to the death of the pilot if the glider encountered an obstacle.

New model

To address these shortcomings Pavel Tsybin designed an improved glider based on the KC-20 featuring a hinged nose for loading and a cockpit mounted well back from the nose above the cargo compartment.

Able to carry 25 troops or 2200kg (4850lbs) of cargo, the new glider, designated Ts-25, was completed during 1944. Too late to see action during the war, the 251 production aircraft served until the mid-1950s.

Kolesnikov-Tsybin KC-20

The KC-20 was larger and rarer than the Gribovsky G-11 and Antonov A-7. Mostly employed to supply partisan units, these gliders flew their missions by night.

Kolesnikov-Tsybin KC-20

Weight (Maximum take-off) 4460kg (9833lbs)
Dimensions Length: 14.12m (46ft 4in), Wingspan: 23.8m (78ft 1in), Height: 2.84m (9ft 4in)
Speed 240km/h (149mph)
Crew 2 plus 20 troops
Armament None

Antonov A-7

A typical glider of its era, the Antonov A-7 could carry six troops and a pilot or up to a tonne of cargo.

From 1931 Oleg Antonov had been chief designer of the Moscow Glider factory that produced thousands of successful sport gliders until 1938, when gliding as a sport was banned and the factory closed following the defection of a glider instructor. In 1940 Antonov's RF-8 design won a competition for a new glider to carry four paratroopers and a pilot. Work had begun on a prototype when advancing German forces caused the evacuation of Antonov and his team to Moscow where work resumed in the former Moscow Glider Factory.

Impaired visibility

The RF-8 was flown for the first time on 2 September 1941 and testing was generally favourable although a worrying tendency for the fuselage to sag when loaded was noted and the windscreen was considered too far from the pilot, impairing visibility especially at night. Despite these shortcomings it was decided to place the RF-8 into production with the demand that its capacity be increased to seven paratroopers plus the pilot. The modified RF-8 was known as the A-7 and retained the same basic dimensions as the earlier craft though tailplane size was reduced and the empty weight of the glider was reduced through redesign of the basic structure.

After a second evacuation as German troops approached Moscow, the A-7 entered production in Siberia. In 1942 the A-7 entered service and was mostly used, in concert with the G-11, to supply partisan forces and transport sabotage groups behind enemy lines. During these operations the tug was usually a Tupolev SB-2 or Ilyushin Il-4, the latter being able to tow two A-7s simultaneously.

Enlarged variant

During 1943, an enlarged variant was developed – the AM-14 – that could carry a maximum of 14 troops. In production from 1944, the AM-14 was used in combat only once when 50 AM-14s transported equipment to partisans near Minsk.

Antonov A-7

'Yellow 7' was on the strength of the 1st Glider Pilots School at Podmosovie in 1943. Famous today for his prolific post-war transport designs, Oleg Antonov's early career was dominated by glider development.

Antonov A-7

Weight (Maximum take-off) 1875kg (4134lbs)

Dimensions Length: 11.5m (37ft 8in), Wingspan: 19m (62ft 2in), Height: 2.53m (8ft 4in)

Speed 300km/h (180mph)

Crew 1 plus 6 troops

Armament None

Yakovlev Yak-6

A light twin that appeared immediately after the beginning of the war, the Yak-6 was mostly used as a liaison aircraft but also operated as a light harassment bomber.

First flown in August 1942, the small Yak-6 was intended from the start to supplement the Polikarpov U-2 biplane as a light utility aircraft and operate as a light nocturnal nuisance bomber. A mainly wooden aircraft powered by two of the M-11 engines that were also used on the U-2 and Yakovlev's own UT-2 trainer, the Yak-6 was intended to be simple to mass produce, although it did boast the relatively advanced feature of retractable undercarriage. The utility version appeared first, seating two crew side by side, and it could carry four passengers or up to 500kg (1102lbs) of cargo that rendered it much more versatile than the U-2 and R-5 biplanes then operating as light transports.

Yak-6NBB

The armed Yak-6NBB, NBB standing for *Nochnoiy Blizhniy Bombardirovshchik* – 'Short-range Night Bomber' – version followed closely on the heels of the basic aircraft. Differing from the standard aircraft in that it sported a dorsal gun position with a rear-facing 7.62mm (0.03in) ShKAS machine-gun and could carry up to five 100kg (220lbs) bombs under the wing centre section, only two Yak-6NBBs were constructed as the aircraft was discovered to have flaws that rendered it ill-suited to nocturnal operation. Handling issues allowed it to spin easily and it was found when flying at night that the dashboard reflected in the windscreen, making it impossible to see anything outside of the aircraft.

Yakovlev Yak-6

Nicknamed the *Dooglasyonok* or 'Baby Douglas', a reference to the C-47, a series of crashes led to the Yak-6's replacement on the production line by the Shche-2. Early aircraft had retractable undercarriage but later machines featured fixed gear, the performance advantage of retracting the wheels being found to be negligible.

Improvised role

The basic Yak-6 was mass produced from November 1942, although only 381 examples were built before production was halted due to the handling issues. Despite this the Yak-6 was widely used, with most VVS units having one on hand for liaison duties and many were modified for action as improvised light night bombers despite the abandonment of the purpose built variant.

Yakovlev Yak-6

Weight (Maximum take-off) 2300kg (5071lbs)
Dimensions Length: 10.35m (33ft 11in), Wingspan: 14.0m (45ft 11in), Height: 2.90m (9ft 6in)
Powerplant Two 100kW (140hp) Shvetsov M-11F 5-cylinder radial engines
Speed 187km/h (116mph)
Range 900km (560 miles)
Ceiling 3380m (11,090ft)
Crew 2
Armament One 7.62mm (0.3in) ShKAS machine gun flexibly mounted in dorsal position; up to 500kg (1102lbs) bombload

Shcherbakov Shche-2

Developed to be a more capacious and versatile transport than the Yak-6 but able to operate from smaller field than the Li-2, the Shche-2 was a highly successful utility aircraft.

Alexei Shcherbakov was a respected engineer who had worked on pressurized cockpits before the war and during 1941 he was in charge of a large repair workshop. Here he identified an urgent need for a cheap light transport aircraft of sufficient size to ferry aircraft engines and other parts for overhaul.

The complexity of developing a completely new aircraft during wartime was not lost on Shcherbakov and the TS-1 was designed to be as simple as possible. It also used a selection of parts already in production for other aircraft, the engines and mountings came straight from the Po-2, the tailwheel was from the Il-2, the shock absorbers from the La-5 and the entire tailplane was a modified Pe-2 unit. The rest of the aircraft was made almost entirely of wood and canvas to minimize use of strategic materials.

Immediate production

The TS-1 prototype was first flown in February 1942. The aircraft demonstrated good takeoff and landing qualities, excellent stability and controllability and was recommended for immediate production. Accordingly the first aircraft – now designated Shche-2 – followed the last Yak-6s down the assembly line at Zavod 47 in October 1943.

Production Shche-2s were of three distinct types: a utility aircraft for up to 14 passengers or cargo, an assault transport for eight-10 paratroopers and a version for training navigators and radio operators.

In service they performed well, popular with crews for their reliability and flying qualities, the Shche-2 was nicknamed *shchuka* – 'pike' – and became well-known for flights behind enemy lines to supply partisan units.

Postwar, Shche-2s were widely used both by the military and as an airliner by Aeroflot.

In total 550 examples of this unpretentious and highly practical machine were built.

Shcherbakov Shche-2

The Shche-2 was developed and built in great haste but proved to be a rugged and useful aircraft, ideally suited to its role. This 1945 example is fitted with Townend rings over the cylinder heads of its M-11 engines typical of later production aircraft.

Shcherbakov Shche-2

Weight (Maximum take-off) 3700kg (8157lbs)

Dimensions Length: 14.27m (46ft 10in), Wingspan: 20.54m (67ft 5in), Height: 3.80m (12ft 6in

Powerplant Two 86kW (115hp) Shvetsov M-11d radial engines

Speed 155km/h (96mph)

Range 980km (610miles)

Ceiling 3000m (9800ft)

Crew 2

Armament None

Curtiss O-52 Owl

The last example of an outdated concept, the Soviet Union was the only nation to take the Curtiss Owl into action.

The last aircraft accepted into the United States Army Air Corps in the 'Observation' category, the O-52 Owl was a two-seat aircraft intended to perform battlefield reconnaissance and artillery spotting duties. Powered by a 450kW (600hp) Pratt & Whitney Wasp and capable of 350km/h (220mph), the somewhat rotund, high-wing O-52 offered its crew an excellent view from its heavily glazed cabin. Unfortunately, by the time the Owl flew in 1940 it had become clear that it was built to an outdated concept, too large to operate closely with ground forces but too slow and vulnerable to modern fighters. In US service the O-52 saw out the war on uneventful antisubmarine patrol and courier duties.

Behind enemy lines

In total 26 O-52s were shipped to the USSR although only 19 arrived. Operating against Finnish and German forces, the Owls primarily flew in its designed role performing observation and correcting artillery fire, although it was occasionally employed to parachute clandestine operatives behind enemy lines.

In the course of its artillery spotting duties, one O-52 managed to score an unlikely victory over an attacking Bf 109 by ramming it – both the German pilot and the O-52 observer managed to escape by parachute, but the pilot P.K. Zhilinsky was killed.

By 1944 the seven surviving Owls were assigned to the 5th Photography Aviation Squadron, based near Moscow and well behind the front line. From here they flew aerial reconnaissance missions in direct support of high command. Two of the Owls survived the war to enter civil service, reflecting well on the durability of the Curtiss aircraft.

Post-war these aircraft flew for a few more years on geological survey work, one of them remaining employed until at least 1952.

Curtiss O-52 Owl

The personal aircraft of the CO of 50 OAFAE, 'White 1' was based in the Moscow Military District as the European conflict drew to a close in May 1945. The O-52 was an ineffectual combat aircraft, yet enjoyed great longevity of service.

Curtiss O-52 Owl

Weight (Maximum take-off) 2433kg (5364lbs)
Dimensions Length: 8m (26ft 4in), Wingspan: 12.4m (40ft 9in), Height: 2.7m (9ft 11in)
Powerplant One 450kW (600hp) Pratt & Whitney R-1340-51 Wasp 9-cylinder air-cooled radial piston engine
Speed 350km/h (220mph)
Range 1127km (700 miles)
Ceiling 5940m (19,488ft)
Crew 2
Armament One 7.62mm (0.3in) Browning machine gun fixed forward-firing in cowl and one 7.62mm (0.3in) Browning machine-gun flexibly mounted in rear cockpit

Douglas C-47

The Douglas C-47 served with virtually every Allied nation during World War II and the Soviet Union was no exception.

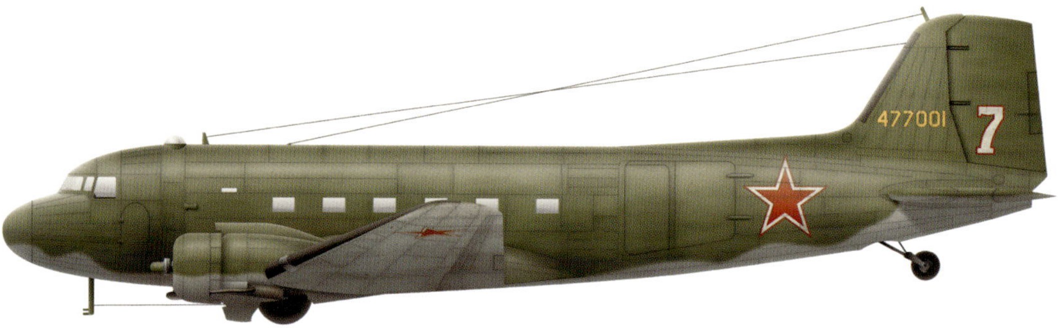

Despite licence building its own Li-2, the USSR also received 707 examples of the C-47, the Douglas-built military transport variant of its influential DC-3 airliner. Fitted with a large cargo door on the left-hand side in place of the passenger door of the DC-3, the C-47 also featured an integral cargo hoist, a strengthened floor and a shortened tail cone to allow a glider towing shackle to be fitted.

The first batch of 25 for the Soviet Union arrived in September 1942.

Stalin's personal aircraft

In service the C-47 was tacitly admitted to be a superior aircraft to the Li-2, as the American instrumentation and equipment was of higher quality and the aircraft was more reliable and better finished. It is no coincidence that the C-47 was the only aircraft Stalin ever flew in, selecting a specially fitted-out C-47 for his journey to and from the Tehran conference of late 1943.

Unlike the Li-2, the C-47 was not equipped with defensive weapons, as US tactical doctrine did not expect to operate transport aircraft without possessing air supremacy and Soviet C-47s suffered from this omission. On operations in Slovakia, for example, C-47s made up a small fraction of the transport fleet but accounted for nearly half of all combat losses. A few aircraft were modified with the UTK-1 turret used in the Li-2, but this was not commonplace.

Likewise the C-47 was never used for bombing missions, remaining almost exclusively a transport aircraft, the exceptions being its occasional use for nocturnal reconnaissance deep behind enemy lines and long range anti-submarine patrol over the Arctic with the Northern Fleet.

Post-war service

After the war, C-47s continued to be used in large numbers – over 100 served Aeroflot until the mid-1950s and many supported Soviet Polar operations. The final Douglas transport to enter Soviet service was a US Navy R3D that was found on an ice floe in 1954. Abandoned two years earlier by US personnel, it was restored and taken into service.

Douglas C-47

This C-47 was assigned to the French-manned Normandie-Niemen squadron. 'White 7' flew with the unit on its return to France in the summer of 1945 with its complement of Yak-3 fighters.

Douglas C-47

Weight (Maximum take-off) 14,061kg (31,000lbs)

Dimensions Length: 19.43m (63ft 9in), Wingspan: 29.11m (95ft 6in), Height: 5.18m (17ft)

Powerplant Two 890kW (1200hp) Pratt & Whitney R-1830-90C Twin Wasp 14-cylinder air-cooled radial piston engines

Speed 360km/h (224mph)

Range 2600km (1600 miles)

Ceiling 8047m (26,400ft)

Crew 4

Armament None

Armstrong Whitworth Albemarle

The Albemarle was a British bomber built of non-strategic materials. It was unpopular and little used in RAF and VVS service.

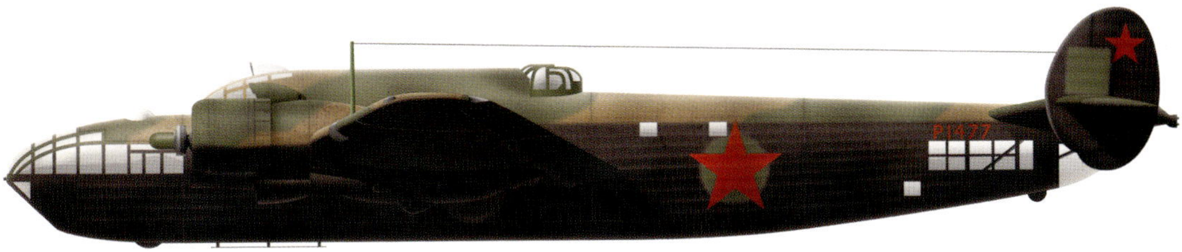

Armstrong Whitworth Albemarle

Still in standard RAF night-bomber camouflage and serial, P1477 was one of the first four Albemarles to arrive in the Soviet Union. This aircraft is shown as it appeared whilst under test with NII V-VS during March and April 1943.

Armstrong Whitworth Albemarle
Weight (Maximum take-off) 16,570kg (36,500lbs)
Dimensions Length: 18.25m (59ft 11in), Wingspan: 23.47m (77ft), Height: 4.75m (15ft 7in)
Powerplant Two 1190kW (1590hp) Bristol Hercules XI 14-cylinder air-cooled radial engines
Speed 426km/h (265mph)
Range 2092km (1300 miles)
Ceiling 5486m (18,000ft)
Crew 4
Armament None

Intended as a back-up to conventional aircraft if aluminium and light alloys fell into short supply, the Albemarle was a creditable attempt to fulfil a need that never eventuated. Had the war progressed in a different way, the Albemarle may have become a significant aircraft, but as things turned out, it operated in its intended role only once and remains obscure.

Paratroop transporter

The first British production aircraft to feature a tricycle undercarriage, the Albemarle was mostly constructed of steel and wood, flying for the first time on 20 March 1940.

In British service the Albemarle operated as a paratroop transport and a glider tug, seeing action from 1943 onwards. The Soviet Union placed a contract for 200 Albemarles in October 1942 and a training unit was set up in Scotland for Soviet crews.

The first aircraft was flown direct to Vnukovo airfield near Moscow on 3 March 1943, followed by a further 12. Testing in the USSR revealed the aircraft was inferior to domestic bombers, so further deliveries were cancelled in favour of C-47s; the remaining handful of Albemarles were formed into the 3rd Regiment of the 1st Transport Division.

Soviet crews were not impressed with the aircraft, as the Albemarle suffered from a protracted takeoff, was sluggish to manoeuvre, maintaining height on one engine was difficult, visibility for the pilot was poor and the cockpit was badly laid out – the flap and undercarriage controls being particularly difficult to operate. The purpose-built Li-2 and C-47s were clearly better transports and able to carry much greater payloads, but the Albemarle was much faster than these aircraft and tended to be used for urgent cargo delivery.

Navy trainers

After a year and a half, the surviving Albemarles were transferred to the Navy and served primarily to train navigators, for this duty six seats were fitted for trainees, each with its own bomb sight. In May 1945 two Albemarles were still in service but by the autumn they had been retired.

SEAPLANES AND MARITIME AIRCRAFT

Soviet Naval Aviation was charged with covering the Soviet Union's large coastline and approaches and it did so in the most challenging conditions, the Northern Fleet operating above the Arctic Circle to escort critical convoys into Murmansk and Arkhangelsk. Its principal aircraft was the wooden but incredibly resilient MBR-2, supplemented by both licence-built and Lend-Lease versions of the Consolidated PBY Catalina.

This chapter includes the following aircraft:

- Tupolev TB-1P
- Shavrov Sh-2
- Tupolev MTB-1
- Tupolev MTB-2
- Beriev MBR-2
- Beriev Be-2
- Beriev Be-4
- Chetverikov MDR-6
- Consolidated PBY Catalina
- Vought OS2U

Finnish soldiers examine a captured Beriev MBR-2 flying boat, February 1943. This one is operating from a frozen water surface on a fixed ski undercarriage.

Tupolev TB-1P

Totally obsolete by 1941, the original TB-1 had first flown as far back as 1925 and was one of the most successful aircraft of the inter-war period.

Designed by Andrei Tupolev, the ANT-4 was simultaneously the first Soviet-designed bomber to enter serial production, the first domestic monoplane bomber and the first Soviet all-metal bomber.

Built at Tupolev's Moscow factory, the ANT-4 was so large that a wall of the building had to be knocked down to allow the aircraft to be removed in sections. The prototype was powered by two 336kW (450hp) Napier Lions but production machines sported two 510kW (684hp) Mikulin M-17s, a licence-built BMW VI. Entering service in 1929 as the TB-1, the aircraft formed the backbone of Soviet long-range bombing capability until gradually supplanted by the more powerful TB-3.

Floatplane conversion

Even before it entered service in land plane form, development of a floatplane conversion of the TB-1 for naval use was underway. Initially utilizing Junkers designed floats that were interchangeable with the standard wheeled landing gear, the first TB-1 floatplane – named 'Land of the Soviets' – used both on its 1929 propaganda flight to New York.

All the floatplanes subsequently produced were conversions of standard TB-1 bombers and were fitted with more advanced floats of British design. From 1932, the TB-1P floatplanes equipped the first Soviet torpedo and mine-laying squadrons operating over the Black Sea, Baltic and Far East.

Obsolete design

By the outbreak of the Great Patriotic War, surviving TB-1Ps were employed purely in the training role, it being clear by this time that the aircraft was obsolete. However, the design lived on as a civil freighter conversion, designated G-1. The final example, operated by Aviaarktika in the far north and polar regions, was eventually retired in 1948.

Tupolev TB-1P

Operating in the summer of 1934 with the 62nd Air Brigade of the Baltic Fleet based in Leningrad, 'Red 4' is seen here with its primary offensive weapon, the TAN-12 torpedo.

Tupolev TB-1P

Weight (Maximum take-off) 6810kg (15,013lbs)

Dimensions Length: 18.90m (62ft), Wingspan: 28.70m (94ft 2in), Height: 5.10m (16ft 9in)

Powerplant Two 510kW (684hp) Mikulin M-17 V-12 liquid-cooled engines

Speed 186km/h (115mph)

Range 950km (590 miles)

Ceiling 4830m (15,846ft)

Crew 6

Armament Two paired 7.62mm (0.3in) PV-1 machine guns flexibly mounted in nose position, two paired 7.62mm (0.3in) PV-1 machine guns flexibly mounted in two staggered dorsal positions; up to 1000kg (2200lbs) of bombs, mines or depth charges, or one 450mm (18in) TAN-12 torpedo

Shavrov Sh-2

The unassuming Shavrov Sh-2 proved extremely useful. Although the basic design dated from 1929, it continued in service until the mid-1960s.

Vadim Shavrov built his first aircraft in a second floor apartment in Leningrad, the wings, hull and tail being lowered to the ground from the balcony. When assembled, the Sh-1 became the first Soviet amphibian to fly, taking to the air on 21 June 1929 powered by a Czech Walter engine. It passed its state tests in Moscow but more power was deemed desirable. The substitution of the M-11 engine necessitated scaling up the design slightly and in this form the aircraft entered production as the Sh-2 with around 300 aircraft being built from 1932–1934.

Ambulance variant

Used as a liaison, utility transport, a trainer and on fishery protection duties and frontier patrol work, the little amphibian's flying qualities were highly praised and the aircraft proved popular. So popular in fact that production was reinstated in 1939 and a further 400 were built. Virtually all pre-war Sh-2s were identical apart from 16 examples of an ambulance variant, the Sh-2S. In this configuration a medic would sit next to the pilot and the patient would be slid on a stretcher into the fuselage, immediately aft of the cockpit, and covered by a removable canopy.

During World War II the VVS and naval aviation operated many Sh-2s primarily for liaison and training duties, and this simple aircraft proved reliable and useful. Its qualities were not missed on its enemies either: two Sh-2s were captured by Finland and pressed into service. One of these examples was used by the commander of the Air Force as his personal aircraft. After the war the Sh-2 reverted to its manifold civilian duties, some aircraft were re-engined in 1951 with a more powerful version of the M-11 engine and in 1952 an enclosed cockpit version was developed. The Sh-2 was widely used in the Caspian Sea and on the rivers and lakes of Siberia until 1964.

Shavrov Sh-2S

Weight (Maximum take-off) 937kg (2066lbs)
Dimensions Length: 8.2m (26ft 11in), Wingspan: 13m (42ft 8in), Height: 2.8m (9ft 2in)
Powerplant One 75kW (101hp) Shvetsov M-11A 5-cylinder air-cooled radial piston engine
Speed 140km/h (87mph)
Range 400km (250 miles)
Ceiling 3500m (11,500ft)
Crew 1 or 2
Armament None

Shavrov Sh-2S

This anonymous example of the Sh-2S operated in the medical evacuation role on the Leningrad Front. Unusually, it is not fitted with the canopy covering the stretcher in the fuselage immediately aft of the cockpit.

Tupolev MTB-1

Despite being a development of a Chetverikov design, the MTB-1 was the only Tupolev flying boat to enter production.

Tupolev MTB-1

Soviet flying boats rarely had exciting colour schemes. This all-silver MTB-1 serving with 124 Naval Heavy Squadron of the Black Sea Fleet in 1939 is devoid of individual markings.

The Chetverikov MDR-3 was a long-range flying boat flown in 1932 that demonstrated poor performance but was noted for its excellent hull which possessed good hydrodynamic qualities. Suspicions about Chetverikov's ability to ameliorate the performance issues meant the aircraft was passed on to the Tupolev design bureau for further development.

Change in designation

Known to its designers as Tupolev ANT-27, it featured an entirely new wing of greater area and a new tail unit mounted on a broadly unchanged hull. A major problem of the MDR-3 was vibration of the tail unit caused by problematic airflow from the tandem engine installations. The arrangement of two tandem pairs of engines was replaced by three 600kW (805hp) Mikulin M-34Rs in an unusual arrangement of one tractor and two pusher units, all strut mounted in separate nacelles above the wing. Construction was complete by 7 March 1934, and the fact that the modifications had resulted in an essentially new aircraft reflected a change in designation to MDR-4.

Development was delayed by the loss of the prototype in a fatal takeoff accident in April 1934, the second prototype taking to the air six months later. In tests, the performance of the new flying boat was found to be an improvement over the MDR-3 but hardly spectacular, although it was noted to have retained the MDR-3's excellent handling on the water and was easy to fly. The urgent need for long-range patrol aircraft took precedence over any reservations regarding its pedestrian performance and production began in 1936, despite the loss of the second prototype in another crash. Production of the aircraft – now designated MTB-1 – was slow due to the factory's inexperience with metal construction, ending after only 15 had been built due to the decision to licence-build the vastly superior Consolidated PBY. The handful of production MTB-1s operated as patrol aircraft until 1942.

Tupolev MTB-1

Weight (Maximum take-off) 16,250kg (35,825lbs)

Dimensions Length: 21.90m (71ft 10in), Wingspan: 39.15m (129ft 3in), Height: 8.76m (28ft 9in)

Powerplant Three 600kW (805hp) Mikulin M-34 V-12 liquid-cooled piston engines

Speed 266km/h (165mph)

Range 2000km (1243 miles)

Ceiling 4300m (14,108ft)

Crew 3

Armament One 7.62mm (0.3in) ShKAS machine gun in bow turret, one 7.62mm (0.3in) ShKAS machine gun in dorsal turret, one 7.62mm (0.3in) ShKAS machine gun in tail turret; up to 500kg (1100lbs) of bombs and depth charges

Tupolev MTB-2

Development of the impressive MTB-2 was curtailed by the outbreak of war, although both prototypes saw active service in the first year of the conflict.

Tupolev MTB-2

The second prototype MTB-2 was fitted with a dorsal turret and was an amphibian – it is pictured here as it appeared before the German invasion. After June 1941 it received a very light application of green, wavy camouflage over the silver.

The first prototype of the Tupolev ANT-44 flew on 19 April 1937, although the first flights were made from a wheeled chassis allowing the aircraft to be tested at Moscow Central airfield; the first takeoff from water was made on the Khimki reservoir during November. A second prototype soon followed and the first was modified into an amphibious aircraft.

Test reports were excellent and it was recommended that the ANT-44 enter production as the MTB-2. Difficulties were encountered finding a suitable production location that would ultimately remain unresolved when war broke out. Meanwhile the second prototype was used to set a string of class records during 1940.

Accident prone

Even before the war, flying the MTB-2 was eventful. During one flight one of the engines caught fire and test pilot Ivan Sukhomlin managed to save the aircraft by rapidly alighting then taxiing at high speed into port where the blaze was extinguished by a fire boat. On another occasion the aircraft made an emergency landing during a storm after running out of fuel – after several hours adrift the flying boat was taken in tow by a fishing boat and brought safely back to base.

Both aircraft were based in Sevastopol when Operation Barbarossa was launched and both spent a busy period on a variety of operations over the Black Sea, including bombing Bucharest on 17 October, repeatedly bombing Ploesti oilfield and taking part in the evacuation of Sevastopol. During bombing raids the crew used the immense endurance of the MTB-2 to repeatedly bomb the same target at intervals over several hours, one bomb at a time (the MTB-2 carrying up to 10) to maximize the psychological effect on the enemy. On 10 August 1942, one of the MTB-2s was destroyed on takeoff by enemy fighters, the ultimate fate of the second is unknown.

Tupolev MTB-2

Weight (Maximum take-off) 21,000kg (46,297lbs)
Dimensions Length: 22.42m (73ft 7in), Wingspan: 36.45m (119ft 7in), Height: 7.88m (25ft 10in)
Powerplant Four 710kW (950hp) Tumansky M-87A 9-cylinder air-cooled radial piston engines
Speed 355km/h (221mph)
Range 1820km (1130 miles)
Ceiling 6500m (21,300ft)
Crew 7–8
Armament One 20mm (0.78in) ShVAK cannon in bow turret, one 20mm (0.78in) ShVAK cannon in tail turret, two 7.62mm (0.3in) ShKAS machine guns in dorsal turret; up to 4000kg (8000lbs) of bombs or depth charges

Beriev MBR-2

The rugged MBR-2 was by far the most numerous flying boat to serve the Soviet Union. It delivered dependable service in all maritime theatres in which Soviet forces operated, from the Arctic to the Pacific.

Beriev MBR-2

With field applied green over the standard pre-war silver, this MBR-2 was serving with 118 MRAP of the Northern Fleet during the summer and autumn of 1941.

Devised on his own initiative by 27-year-old Georgy Beriev during 1930, the proposal of the MBR-2 was presented at a meeting of senior Soviet designers and recommended for development. The all-wood prototype took off for the first time from the Black Sea on 30 April 1932. Powered by a BMW VI, the performance of the MBR-2 was not exactly sparkling but the aircraft was strong and possessed good handling and pleasant takeoff and landing characteristics.

The aircraft entered production in 1934 powered by the M-17, a licence-built BMW VI, and it could carry 300kg (661lbs) of bombs and was defended by a 7.62mm (0.3in) PV-1 machine gun in the bow and another amidships. About 300 of the M-17-powered aircraft were built before 1935 when production switched to an improved version featuring an enclosed cockpit, a quick firing 7.62mm ShKAS in a dorsal turret and a Mikulin M-34 engine offering nearly 100kW (134hp) more power. A civil airliner variant called the MP-1 – or MP-1bis when fitted with the M-34 – seating six passengers in an enclosed cabin was developed concurrently with the naval version. Production of the M-34-powered MBR-2 continued until 1940, although the service life of the aircraft would continue for many years.

The first examples of Beriev's flying boat entered service during 1934, replacing Dornier Wals and Savoia Marchetti SM.62s. By 1937 the MBR-2 was the most widespread maritime aircraft in Soviet Naval units, a position it would maintain for some years despite its increasingly manifest obsolescence. The 1939–1940 Winter War with Finland saw the MBR-2's first use on operations and the aircraft acquitted itself well in harsh conditions, flying reconnaissance missions as far north as the Barents Sea and attacking Finnish shore installations and shipping. During this conflict the MBR-2 operated on fixed skis from the frozen surface of Lake Ladoga.

The scarcity of Finnish fighter aircraft had allowed the MBR-2 to operate largely unmolested, but during the Great Patriotic War conditions were different and the aircraft's extreme vulnerability to fighters became apparent. The Baltic Fleet lost two-thirds of its Berievs during 1941, and it was used more often by night if enemy fighter activity was likely. Operations in the Black sea tended to be beyond the range of fighters and the MBR-2 maintained its designed role of naval reconnaissance and anti-shipping missions. The last combat missions by the MBR-2s were flown in the Pacific against Japanese shipping.

Beriev MBR-2

Weight (Maximum take-off) 4245kg (9359lbs)
Dimensions Length: 13.50m (44ft 3in), Wingspan: 19.00m (62ft 4in), Height: 5.36m (14ft 5in)
Powerplant One 559kW (750hp) Mikulin AM-34N V-12 liquid-cooled piston engine
Speed 234km/h (145mph)
Range 690km (716 miles)
Ceiling 7400m (24,278ft)
Crew 4–5
Armament One 7.62mm (0.3in) PV-1 machine gun flexibly mounted in bow, one 7.62mm (0.3in) ShKAS machine gun in dorsal turret; up to 300kg (661lb) bombload

Beriev Be-2

The first Soviet seaplane designed for shipborne catapult operations, the Beriev Be-2 suffered from handling problems that were never fully overcome.

The first catapult for aircraft launch in the Soviet Navy was fitted to the battleship *Parizhskaya Kommuna* in 1930. It was one of two KR-1 catapults supplied by the German firm Heinkel along with 15 examples of the KR-1 scout biplane flying boat. The KR-1 was never an entirely satisfactory aircraft and a replacement was eagerly sought. To address this need Georgy Beriev designed the KOR-1, but unfortunately this aircraft never proved satisfactory either. In the absence of any readily available alternative, 12 were built and the aircraft saw brief but intense action in the first six months of war.

Metal construction

First flown in April 1936 the all-metal KOR-1 was fitted with a Shvetsov M-25 engine endowing it with the less than stellar top speed of 277km/h (172mph). Worse still, its handling in the air was sluggish and on the water it was poor, although when fitted with a wheeled undercarriage in place of the float the performance was somewhat improved. Persistent overheating and corrosion of the internal structure caused by water ingress also dogged the little Beriev. Even the wing-folding procedure was inadequate, requiring an excessive 40 minutes to fold or unfold the mainplanes. Further modifications improved the aircraft but its performance was still much criticized. Eventually the KOR-1 made its first test flights from a catapult on board the cruiser *Voroshilov* during April 1940.

In late June 1941 the KOR-1s were re-designated Be-2 and went into action – in the Baltic it was used for reconnaissance, one Be-2 being captured by the Finns but with the Black Sea Fleet, however the Be-2 was employed in combat. Operating from land with the optional wheeled undercarriage rather than as a floatplane, the five Be-2s went into action as light bombers in the defence of Sevastopol, attacking the advancing German army until the final example was irreparably damaged by artillery fire on 3 December 1941.

Beriev Be-2

Weight (Maximum take-off) 2626kg (5789lbs)

Dimensions Length: 8.88m (29ft 1in), Wingspan: 11.00m (36ft 1in), Height: 4.42m (14ft 6in)

Powerplant One 522kW (700hp) Shvetsov M-25A nine-cylinder, air-cooled, radial piston engine

Speed 277km/h (172mph)

Range 530km (329 miles)

Ceiling 6600m (21,654ft)

Crew 2

Armament Two 7.62mm (0.3in) ShKAS machine guns fixed forward-firing in fairings on upper wing centre section, one 7.62mm (0.3in) ShKAS machine gun flexibly mounted in rear cockpit; up to 100kg (220lbs) bombload

Beriev Be-2

KOR-1 'Red 5' of an unknown unit at Krasnoyarsk as it appeared in the late 1930s. Most combat missions undertaken by the type were from land on a wheeled undercarriage.

Beriev Be-4

The Be-4 was an excellent aircraft that finally gave the Soviet Navy the shipboard scout it had wanted since 1930. Despite this, only 46 were built.

Seeking to replace the disappointing KOR-1/Be-2 and restore his somewhat compromised reputation, Georgy Beriev started to develop a new catapult-launched aircraft in 1939. This aircraft represented a complete departure from the design of its predecessor, as it was a monoplane flying boat rather than a biplane floatplane and benefitted from a considerably more powerful 746kW (1000hp) Shvetsov M-62 engine. On 21 October 1940 the first prototype, initially known as the KOR-2, made its maiden flight. Flight characteristics were excellent, the aircraft was easy to fly and it demonstrated good seaworthiness. Renamed the Be-4 in December 1940, production was expected to commence in 1941.

Black Sea fleet

Unfortunately production was delayed first by the inexperience of the factory assigned to build the Be-4 and then by the movement of the factory to two different locations within two years, meaning that only a paltry amount of aircraft were available from the summer of 1942 onwards. Initial use was with the Black Sea fleet, although possessing no catapult equipped ships at this time the Be-4s were employed on anti-submarine warfare (ASW) patrol. During this period German Do 24 and Bv 138 flying boats were encountered but no engagements took place.

Polar aviation service

A catapult was eventually fitted to the cruiser *Molotov* and the Be-4 began its intended role in October 1944. By this time the aircraft had also seen service over the Baltic and in the Arctic, where it would experience its sole combat use: on 28 August 1943, a Be-4 bombed a German submarine that had sunk a Soviet transport, although it is not known if any damage was inflicted. Surprisingly the Be-4 in question was technically a civil aircraft, having been transferred to the Polar aviation service.

Post-war the Be-4's career was short as the advent of radar and the helicopter rendered this class of aircraft obsolete, and all catapults were removed from Soviet ships during 1947.

Beriev Be-4

The Be-4 wasn't available when it was most sorely required and as a result saw very little combat action. This typical example was operational in 1944.

Beriev Be-4

Weight (Maximum take-off) 3468kg (7646lbs)

Dimensions Length: 10.50m (34ft 5in), Wingspan: 12.00m (39ft 4in), Height: 4.05m (13ft 3in)

Powerplant One 746kW (1000hp) Shvetsov M-62 nine-cylinder, air-cooled, radial piston engine

Speed 356km/h (221mph)

Range 1150km (716 miles)

Ceiling 8100m (26,575ft)

Crew 3

Armament One 7.62mm (0.3in) ShKAS machine gun flexibly mounted in rear cockpit; up to 300kg (661lbs) bombs or depth charges

Chetverikov MDR-6

The only Chetverikov design to enter production, the MDR-6 was a modern all-metal flying boat of stressed skin construction but was built in trivial numbers.

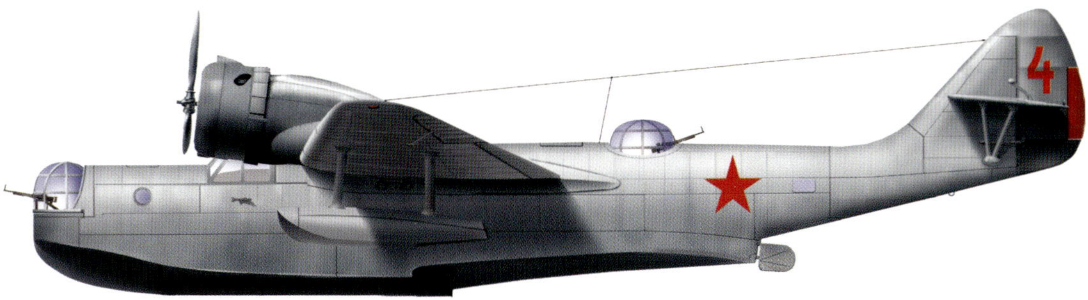

Intended as a replacement for the Beriev MBR-2, Igor Chetverikov's elegantly designed gull-winged MDR-6 – MDR for *Morskoi Dalnii Razvedchik* or 'long-range naval reconnaissance' – first flew during July 1937. It was judged superior to Beriev's competing MDR-5 and was approved for serial production in 1938. Production aircraft were fitted with two of the new Shvetsov M-63 engines, offering 821kW (1100hp) for takeoff replacing the 520kW (697hp) M-25s of the prototype and performance was significantly improved, maximum speed being 360km/h (224mph).

Defensive armament was weak, consisting of just two machine guns, a 12.7mm (0.5in) Berezin UBT in the dorsal turret and a 7.62mm (0.3in) ShKAS in the nose but underwing racks allowed up to 1000kg (2205lbs) of bombs to be carried – more than twice as much as the MBR-2.

Production of the MDR-6 began in 1940 in Taganrog in southwestern Russia, the aircraft being re-designated Che-2 in December 1940, but only 20 aircraft were built before the approach of German forces necessitated abandoning the factory and no further examples were constructed.

Chetverikov MDR-6

This MDR-6 was operating over the Baltic when war broke out and subsequently received a disruptive camouflage of green sprayed over the fuselage sides.

Baltic service

Che-2s entered service in the Baltic at the end of 1940 and the aircraft was later used by both the Black Sea and Northern Fleets. Its good performance and excellent water handling qualities were much appreciated by crews, although they were less keen on the front turret's propensity to take on water during takeoff. The Che-2's service life was destined to be brief however, as structural problems on the production aircraft caused all to be withdrawn during 1942.

Chetverikov designed and built a series of ever-improved developments of the Che-2 between 1942 and 1945, but despite good flight test results none went into production.

Chetverikov MDR-6

Weight (Maximum take-off) 7200kg (15,873lbs)
Dimensions Length: 15.73m (51ft 7.25in), Wingspan: 19.4m (63ft 7.75in), Height: 4.30m (14ft 1in)
Powerplant Two 821kW (1100hp) Shvetsov M-63 nine-cylinder, air-cooled, radial piston engine
Speed 360km/h (224mph)
Range 2650km (1647 miles)
Ceiling 9000m (29,530ft)
Crew 3
Armament One 7.62mm (0.3in) ShKAS machine gun in bow turret, one 12.7mm (0.5in) Berezin UBT machine gun in dorsal turret; up to 1000kg (2205lbs) of bombs or depth charges

Consolidated PBY Catalina

Built under licence as the GST, the Soviet Union also received 185 PBY flying boats via Lend-Lease arrangements.

The first Catalina in Soviet hands was secondhand. Known as the 'Guba', it was the first civil PBY built by Consolidated in 1937, which the Soviets asked to buy from owner Dr. Richard Archbold in order to search for Sigismund Levanevsky and his crew who had disappeared during a flight across the North Pole. Although Levanevsky was never found, Guba made a very positive impression and three military PBYs were ordered and a production licence obtained. Meanwhile, Guba was modified into an armed maritime patrol aircraft and after a busy early war period, including searching for survivors from Convoy PQ-17, she was sunk by shellfire while at anchor on 25 July 1942.

Coastal patrol

The three PBYs delivered into pre-war Soviet service differed from all other US-built examples in that they were powered by Wright Cyclone engines. Although the M-62 was a Soviet-built Cyclone, oddly the licence-built PBYs were powered by two Tumansky M-87s (a derivative of the French Gnome et Rhône 14K *Mistral Major*). Designated the GST (for *Gydro Samoliot Transportnyi* or 'Hydro Aircraft Transport'), 27 were delivered by June 1941 and used for coastal patrol. Bizarrely, in November 1941 a GST was flown to Cyprus by a Soviet sailor and was taken into RAF service with the serial HK850.

In addition to the Russian-built GST, the Soviet Navy received 48 PBY-6 amphibians and 137 examples of the PBN Nomad, which was basically a PBY-5 built by the Naval Aircraft Factory with a more efficient re-shaped extended hull and a greatly increased fuel capacity, allowing for a 50 per cent increase in range. Delivered between February 1943 and March 1945 and designated KM-1 in the USSR, many were later re-engined with the ASh-82FN and redesigned KM-2, serving until the mid-1950s. The Nomads and PBYs flew in all theatres in which the Soviet Navy operated from the Arctic to the Pacific and contributed to the sinking of at least five U-boats.

PBN-1 Nomad

The majority of Nomads built went to the Soviet Union. 'White 30' was finished in the standard US Navy tricolour scheme and is pictured as it appeared at Fairbanks, Alaska, on 30 September 1944 during its ferry flight to the USSR.

PBN-1 Nomad

Weight (Maximum take-off) 16,500kg (36,376lbs)

Dimensions Length: 19.72m (64ft 8in), Wingspan: 31.80m (104ft 4in), Height: 6.48m (21ft 3in)

Powerplant Two 890kW (1200hp) Pratt & Whitney R-1830-92 Twin Wasp 14-cylinder air-cooled radial piston engines

Speed 300km/h (186mph)

Range 5955km (3700 miles)

Ceiling 4600m (15100ft)

Crew 5–7

Armament Two 7.62mm (0.3in) Browning M2-AN machine guns in bow turret, one 7.62mm (0.3in) Browning M2-AN machine gun flexibly mounted in ventral hatch, one 12.7mm (0.5in) Browning M2 machine gun flexibly mounted in each waist blister; up to 1814kg (4000lbs) of bombs or depth charges

Vought OS2U

The Vought OS2U Kingfisher had a long and successful career in US service but only two operated for the USSR, serving from 1944 to 1949.

Vought OS2U

The two Kingfishers operated by the Soviet Navy were imaginatively coded '1' and '2' during their service on *Murmansk*. Both remained in their standard US scheme of grey over white for the duration of their time in Russian waters.

The Soviet Union's two Vought Kingfishers arrived in an unusual manner. After the surrender of Italy to the Allied powers in June 1943 it was expected that the vessels of its modern navy would be split between the US and UK, the UK in particular being keen to operate Italian battleships against the Japanese. Stalin, however, had other ideas, demanding a third of the fleet despite Soviet forces having played no part in the Mediterranean campaign. With the Italian government of General Badoglio (a staunch Fascist) loath to hand over vessels to the Soviet Union – and the UK and US keen to keep Stalin happy in the lead up to the Normandy landings – it was agreed that several British and American ships would be loaned to the USSR through Lend-Lease channels in lieu of the Italian ships.

The British loaned a battleship, eight destroyers and four submarines, but the United States supplied only one vessel, USS *Milwaukee*, an Omaha Class cruiser. *Milwaukee*, renamed *Murmansk*, joined the Soviet Northern Fleet in Murmansk on 20 April 1944 complete with her complement of two Vought OS2U-3 Kingfishers.

Training role

The two floatplanes do not appear to have been used in combat, *Murmansk* spending most of the remainder of the war at anchor in Arkhangelsk, but they were used for a variety of auxiliary purposes, not least training pilots and crew in difficult catapult-launching operations.

Vought OS2U

Weight (Maximum take-off) 2722kg (6000lbs)

Dimensions Length: 10.24m (33ft 7in), Wingspan: 10.94m (35ft 11in), Height: 4.47m (14ft 8in)

Powerplant One 340kW (450hp) Pratt & Whitney R-985-AN2 Wasp Junior nine-cylinder air-cooled radial piston engine

Speed 275km/h (171mph)

Range 1461km (908 miles)

Ceiling 5500m (18,200ft)

Crew 2

Armament One 7.62mm (0.3in) Browning M2-AN machine gun fixed forward-firing in upper cowling, one 7.62mm (0.3in) Browning M2-AN machine gun flexibly mounted in rear cockpit; up to 295kg (650lbs) bombload

Murmansk sailed back to the US in 1949 with both Kingfishers on her catapults and was formally returned to the US Navy on 16 March as USS *Milwaukee* once again, thus ending the uneventful career of the rarest Lend-Lease aircraft to be operated by Soviet forces.

Glossary

ADD	*Aviatsiya Dal'nevo Deistviya*	Long Range Aviation Arm
APDD	*Avia Polk Dahl'nevo Deystviya*	Long Range Aviation Regiment
Aviaarktika		The independent Arctic directorate of Aeroflot
BAD	*Bombardirovochny Aviatsionny diviziya*	Bomber Aviation Division
BAP	*Bombardirovochny Aviatsionny Polk*	Bomber Aviation Regiment
GvIAP	*Gvardeyskiy Istrebitelny Aviatsionny Polk*	Guards Fighter Aviation Regiment
GvShAP	*Gvardeyskiy Shturmovoy Avialsionny Polk*	Guards Attack Aviation Regiment
IAP	*Istrebitelny Aviatsionny Polk*	Fighter Aviation Regiment
MRAP	*Morskaya Razvedyvatelny Aviatsionny Polk*	Naval Reconnaissance Aviation Regiment
MTAP	*Minno-Torpednyi Aviatsionnyi Polk*	Minelaying-Torpedo Aviation Regiment
NII GVF	*Nauchno-Issledovatelskiy Institut Grazdahnskovo Vozdooshnovo Flota*	Civil Air Fleet Scientific Research Institute
NII-VVS	*Nauchno Issledovatelyskiy Institut – Voyenno Vozdushnye Sily*	Scientific and Research Institute of the Military Air Forces
OKB	*Opytno Konstrooktorskoye Byuro*	Experimental Design/Construction Bureau
PVO	*Protivovozdushnoy Oborona*	Air Defence Forces
RAP	*Razvedyvatelny Aviatsionny Polk*	Reconnaissance Aviation Regiment
ShAP	*Shturmovoy Avialsionny Polk*	Attack Aviation Regiment
Shturmovik		Attack aircraft
TsAGI	*Tsentralniy Aerogidrodinamicheskiy Institut*	Central Aerohydrodynamic Institute
TsKB	*Tsentral'noye Konstruktorskoye Byuro*	Central Construction Bureau
UAP	*Oochebny AviaPolk*	Instructional Regiment
VVS	*Voyenno Vozdushnye Sily*	Military Air Forces
ZAP	*Zapasnoy Aviatsionny Polk*	Reserve Aviation Regiment
Zavod		Plant/Factory

Index

Page numbers in **bold** refer to illustration captions.

Picture Credits

PHOTOGRAPHS:

AirSeaLand.Photos: 12, 21, 22, 32, 52, 56, 92
Amber Books/Aerospace: 6, 7, 26, 28, 39, 42, 73, 78, 82
Courtesy of the Central Museum of the Armed Forces, Moscow via Stavka: 8, 31, 47, 61, 63
Library of Congress: 51
From the fonds of the RGAKFD in Krasnogorsk via Stavka: 102

ARTWORKS:

Amber Books/Aerospace: 10–18, 21–23, 25–29, 32–33, 35, 36–37, 40–45, 48–49, 54, 62, 72, 74–75, 77–83, 86–87, 88, 90
David Bocquelet: 64, 94–95, 99, 114–123
Vincent Bourguignon: 103
Ed Jackson (artbyedo.com): 5, 20, 24, 30–31, 34, 38–39, 46, 50–51, 53, 55, 58–60, 66–71, 76, 84–85, 89, 91, 96–98, 100–101, 104–111